3-00

The Green Holiday Guide

Spain &
Portugal

Cottages and campsites on organic farms
Environment-friendly guesthouses and B&Bs

GW00356835

ECEAT

Acknowledgements

This Green Holiday Guide has come about with great help, moral support and encouragement from a variety of people and organisations. I would especially like to thank Ben, Miek and Maria Schasfoort, Christina Hallström, Misha Hoyinck, Isel, Severino García González, Vidal Postigo Escribano, Anton Weijts, Mirthe Bos, Clara Goudsbloem, Elena Parini, Alice Ruijzenaars, Geert Snoeijer and Ernst Schilp for their contributions and John Elford for his quick and thorough editing.

The photographs of places to stay and other illustrations have been supplied by the hosts, ECEAT, Ben Schasfoort, the Spanish Tourist Board and ICEP-Portugal

Coordination and research: ECEAT-NL, Esther Schasfoort
Research Spain: Jeanne van Sebille, Frank Lojacono, Maxim Holtel
Research Portugal: ECEAT-PT, Mirjam Olsthoorn, Adelaide Almeida
Graphic design and layout: ECEAT-NL, Ernst Schilp
Editing: Green Books, John Elford
Translations Spain & Portugal: Word's Worth, Utrecht Tel: 030 280 16 60

Published in April 2002 by
Green Books Ltd, Foxhole, Dartington,
Totnes, Devon TQ9 6EB
sales@greenbooks.co.uk
www.greenbooks.co.uk

with the
European Centre for Ecological and Agricultural Tourism (ECEAT)
P.O. Box 10899
1001 EW Amsterdam
The Netherlands
eceat@antenna.nl
www.eceat.nl

© Copyright ECEAT-NL 2002
Design Ernst Schilp
Printed in the Netherlands by
W.C. Den Ouden BV, Amsterdam

Printed on non-chlorine bleached paper, EMAS (Eco Management Audit Scheme) certified

ISBN: see back cover

Contents

Left: Breakfast in the farmyard, Malberg, Germany; Middle left: grass roof studio at Middle Wood, Lancashire, England; Middle right: horseriding in Umbria, Italy; Bottom left: Posada del Valle, Asturias, Spain; Bottom right: children playing behind the barn, Poland.

Molezon, Cevennes, France

Holidays in Europe's countryside

The European Centre for Ecological and Agricultural Tourism (ECEAT) started back in 1992, when our first initiatives were to help Polish farms to focus on taking holiday guests, and environment-friendly accommodation management.

Since then, hundreds of thousands of families with children, young couples, pensioners and backpackers have discovered one or more of their favourite holiday destinations on organic farms from Lapland to Sicily, and from Donegal to Transylvania. Nowadays, holidaymakers can choose from a selection of some 1500 organic farms and environment-friendly places to stay in areas of natural beauty in Europe's countryside.

The key words for the ECEAT programme are small-scale, informal, environment-friendly, and beautiful natural surroundings. Most of the places to stay in this guide are an ideal base for walking, cycling, riding, bird-watching and water sports, and many have their own programmes on offer, such as painting courses, music-making, health treatments, Tai Ch'i, permaculture, cheesemaking and plenty more.

Let's have a closer look at a couple of ECEAT farms around Europe:

Gabriel Vazquez and Puri Adrian run an organic farm in a remote, lush Basque valley in the north of Spain. The old farmstead is situated in a hamlet, tucked in between two nature reserves. There are numerous footpaths across wonderful hay meadows that are grazed by the indigenous Latxa sheep and teeming with wild flowers, butterflies and birds. Gabriel is a botanist and kiwi grower. The splendour of these natural surroundings inspired him to write two books: *Medicinal Plants of the Basque Country* and *The Magical Power of Plants.* In a room filled with the scents of fresh and dried herbs, Gabriel runs courses in  medicinal botany, cosmetics, permaculture and ecology, and takes groups on botanical excursions. In summer you will find Gabriel at farmers' markets which sell local organic produce and old Basque arts and crafts.

At Castle Creavie in Kirkcudbright, **Charlie and Elaine Wannop** look after 120 hectares of land in the beautiful south-west of Scotland. "Each season brings with it its own work. The skills you will see in everyday use when you visit are centuries old. The training and working of the

Scottish Border Collie dogs who move, herd and catch sheep - truly a wonderful experience. The mosaic of the landscape is defined by the drystone walls - artistic sculptures in themselves, which need regular attention if they are to stand for a hundred years or more. A good wall needs a firm foundation and a solid 'heart'. The building or repairing of a wall can be like a meditation, with the sound of stone on stone, the fitting of the 'jigsaw', the wind blowing in your hair, and the knowledge that what you are doing will stand for another hundred years. It will shelter lambs in a wet April, protect the animals from bad weather, and keep them where they should be!

Not so easy to control with stone walls are our honeybees! The influences of the sun and stars move these spiritual little creatures in a dream-like world of their own; based on the earth, but called from the heavens. Working with the bees is truly an enlightening insight into total self-lessness, caring for the community and the well-being of the whole.

To enhance wildlife, we actively manage our hedgerows and woodlands. 'Laying' the hedges, thinning the woodlands, planting more trees and creating other wildlife habitats on the farm are other seasonal jobs on Castle Creavie."

Elaine Wannop, Co-ordinator for ECEAT in Scotland

On Tisenö Farm on a private island in the Tisnaren lake, central Sweden, **Niklas and Isabelle** run a bio-dynamic farm which has 125 hectares of cultivable land and pastures, and 500 hectares of water for fishing. "Our principal activities are animal husbandry and forestry. A couple of years ago we started building a clay-and-straw house, which is used as a centre for wilderness survival programmes, guided nature tours and creative courses on clay-and-straw house building, cooking with local farm products, scythe techniques and the maintenance of pastures. We have comfortable chalets and a natural campsite by the lakeshore, set in a beautiful wilderness with panoramic lakes and dark pinewoods."

Niklas Palmcrantz, co-ordinator for ECEAT in Sweden

On the 70 hectare La Cerqua estate (a former monastery) in Umbria, Italy, **Gino and Silvana Martinelli** offer full board with tasty and varied wholefood menus (subtly finished with a drop of local truffle oil), wines and liqueurs, all home-made and organic. The estate is carefully managed by a private trust. Owing to its secluded position in this relatively remote part of Umbria, the valley has a very particular microclimate, and as a result there are rich and varied wildlife habitats. Gino has set out a nature trail through the Wizard's Garden, beautifully illustrated by posters at points of interest, which explain the ecosystem around you - for example an oak tree or a marshy pond. He also leads a monthly full moon excursion, in which the secrets of the night reveal themselves.

Willy and Bram Borst at the Buitenplaats in Eenigenburg in the Netherlands run an organic horticultural farm producing flower bulbs. For economic reasons, conventional farmers plant a tremendous number of bulbs on their fields, rendering them vulnerable to all sorts of fungi and diseases, and leading to the use of vast amounts of pesticides. Most consumers are unaware that conventional horticulture is one of the greatest producers of agricultural pollution. Organic farmers have to tackle this in a creative manner. Bram and Willy plant their bulbs as far apart as possible, so enough wind can blow around them, but of course this leads to reduced crops and higher prices. Whereas conventional farmers spray a thin layer of chloride to prevent weeds, Willy and Bram place their bulbs on a bridge of sand, enabling them to carefully hoe the soil in spring, using good old manual labour. In spring, the young flower heads are cut off in order to prevent the bulbs from producing seeds, which would be detrimental to the bulb. After cutting, the bulbs mature and are reaped by the end of June. Conventional farms prolong the time that bulbs spend underground, and spray with artificial fertilizer, producing larger bulbs. After reaping, the bulbs must be quickly and carefully dried, to guard against more potential disease. Then the peeling, sorting and counting can begin, and everything prepared for the sale.

Photo: Wendy van der Does

If you are interested in farm work in exchange for board and lodging, have a look at page 158.

Many of these places are also members of World Wide Opportunties on Organic Farms, WWOOF (see also pages 111 and 153): www.wwoof.org

ECEAT is a non-profit European association according to the law of the Netherlands. If you are interested in finding out more about ECEAT, its Green Holiday Guides to other European destinations, or in supporting us, contact us at
ECEAT
P.O. Box 10899
1001 EW Amsterdam
The Netherlands
Tel: 00 31 20 668 10 30
Fax: 00 31 20 463 05 94
eceat@antenna.nl
www.eceat.nl

A full list of ECEAT contacts is given on page 164.

WWOOF

Organic farming and environmental protection in Europe

Untreated farmland attracts more species, both in terms of quantity and variety - animals, plants, herbs and wild flowers - than conventional fields where pesticides and artificial fertilisers are used. Together with national parks and other protected habitats, organic farms play a key role in weaving a patchwork of sound ecological areas which enable indigenous inhabitants to survive and thrive, bringing back the authentic character and natural balance to our European countryside.

Various crises in agriculture have wreaked havoc in the farming community in Europe in recent years, and

Organic farms are an attractive environment for wildlife and flowers

have helped to create an emerging political consensus that things cannot go on in the same way. Post-World War II, agricultural policy has gone beyond its goals of rebuilding Europe and producing sufficient food. There is of course a conflict between a policy of protecting nature reserves, and the promotion of production-linked incentives for farming activities.

Increasingly, farmers are involved in land and nature management: the current huge overcapacity in agricultural production will undoubtedly be reduced, forcing farmers to switch to alternatives. Tourism, and the protection and management of nature, will play an important role in securing farmers' incomes and keeping small-scale rural communities alive and authentic. Nature's riches are a 'luxury good', and nowadays have a market value. If exercised in a sustainable manner, and provided revenues are allocated to those who help to conserve nature, tourism can help rural communities to survive economically in the long term. By enjoying a wonderful holiday in the countryside we help to conserve nature's riches, and keep our environment beautiful.

How to use this guide

This guide lists 116 places to stay in Spain and 48 places to stay in Portugal. The numbers on the map on page 10 correspond with the places to stay in the chapters for Spain and Portugal respectively. An easy reference chart of all places to stay, classified by type of accommodation (campsite, lodging, self-catering and hostel) and on the spot services like swimming, bike rental, meals etc, can be found on page 158.

The place to stay you find in this guide are on certified and uncertified organic farms, non-working farms, small-holdings, ecologically-run campsites and hostels. In many cases food comes from the owners' vegetable garden and orchard, and is produced for the sole use of the owners and their guests. Wherever food is served, as much of it as possible will be organic and locally sourced.

A few places have installed facilities specifically geared to wheelchair-users, ranging from simple ones like ramps in the house and outside, to special facilities in rooms and bathrooms. If so, this is also mentioned in the reference chart on page 158.

Keys to symbols used and prices

pp: per person	1p: single room
pn: per night	2p: double room
pw: per week	2p: double room
(ls): low season	4p: quadruple room
(hs): high season	B!: booking essential

Generally, prices for camping are per night, lodging and B&B are given per room per night, self-catering per week, and hostel and bunk houses either per person or for the whole place. Meals are priced per person.

The prices listed in this guide are indications only, and aren't comprehensive. Besides camping prices which are mentioned in the guide, additional fees may be asked for parking cars and other vehicles, for the use of electricity, for hot showers and if you bring pets. At some places, the use of kitchen is offered; usually a small fee is asked for this service. High season rates are to be expected from July to September, at Christmas and at Easter, but there are exceptions to these dates. Make sure to inform yourself when booking.

Several campsites in Portugal use general ECEAT-prices that can be found on page 115. If so, a reference to the page is made in the entry.

Reservations and currency

Especially for lodging, self-catering and hostels, it is recommended that you book in advance throughout the year. In season, this also goes for campsites. When booking is essential you will find the B! mark.

Reservations from abroad are made by first dialling 00 34 for Spain and 00 351 for Portugal. When booking ahead the owner will ask for a down payment, ranging from 20% to 50% of the costs of your total stay.

Prices are given in Euro (€). At the time of printing one Euro equals approximately £0.62 and US $0.87. You can check www.xe.com/ucc/ for the latest conversion rates.

More organic places to stay

http://www.organicholidays.com will ultimately be a guide to organic places to stay, to eat and to visit. http://www.organic-placestostay.com is the first of these guides, and lists bed and breakfasts, guest houses and small hotels where organic and local produce is used according to availability. Also listed is holiday accommodation to rent on organic farms and organic smallholdings. There are about 800 places to stay around the world. Organic Places to Stay is by no means complete - there are many more places to add to the website, so keep checking as new places are added every week.

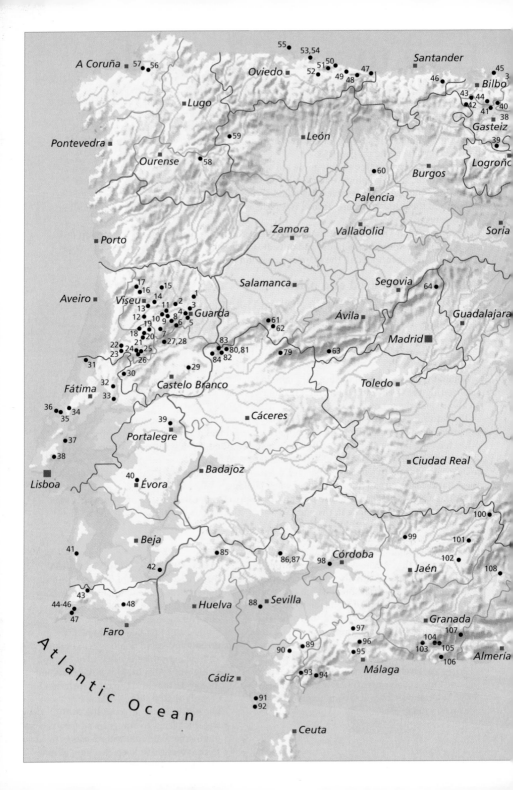

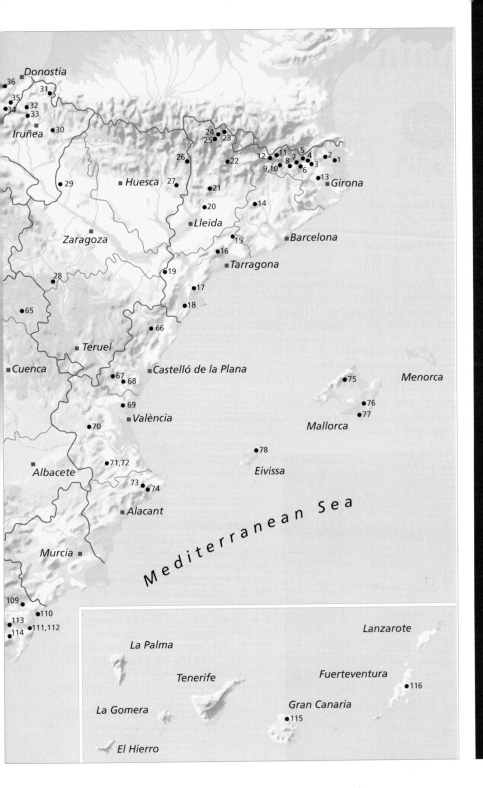

Spain

€SPAÑA

The kingdom of Spain takes up the lion's share of the Iberian Peninsula. Many people think of Spain as a land of sun, sea and blossoming citrus trees, but there is much more. This vast country - including the Balearic and Canary Islands, covering some 500,000 square kilometres - has a wide variety of vegetation, landscapes, climates and architectural styles.

Spain has large plateaus and mountain ranges that stretch across the entire country, as well as regional features: fertile river valleys, arid plains, rolling hills, rocky coastlines and sandy beaches. The climate in the interior is continental, while temperate ocean air dominates in the north and Mediterranean winds warm the *Costas*.

Through the centuries, vastly different cultures have left their mark on Spain. The Etruscans arrived in the east, from across the Mediterranean. The north-west was colonized by Celts, who introduced the *gaita* (bagpipes). The Moors came from the south and settled throughout the Iberian Peninsula. The resulting differences in the local folklore traditions and cuisine can be observed during village festivals such as Carnival, which is celebrated in February.

Spain is excellent for walking and cycling holidays. It has many walking trails designated either as GR (*Grandes Rutas,* long-distance trails) or PR (*Pequeñas Rutas,* local trails). There are also paths alongside, or on, abandoned railways. Walkers may also take the *Vías Pecuarias* - a network of ancient local and regional shepherding routes with a total length of 125,000 km. The longest of these routes are the *Cañadas Reales,* which were used to herd sheep from arid southern Spain to the green north and back again each year.

'Regular' and Organic Agriculture

Spain remained an agricultural society longer than most other Western European countries. In 1950, half the workforce was still employed in agriculture. By 1993, it was only 10 percent.

Because of the variations in Spain's climate and topography, the country's agriculture differs by region. Some three-quarters of the farms in Spain are smaller than 10 ha in area. Large farms are found in Extremadura and Andalucia. Here, bulls and pigs graze great expanses of land dotted with cork oaks. Many extensive sheep farms are found north of Extremadura, around Salamanca and on the Castillian plateau. The rainy and mountainous provinces on the northern coast specialize in small-scale cattle farms. Crop farming and viniculture are concentrated in northern and central Spain, while the warmer Mediterranean coast (Valencia and Murcia) is used for market gardening, citrus, almond, olive and fig groves, and rice growing. The mountain slopes of the southern Andalucian coast also feature vineyards. Their autumn colours are magnificent, while in February almond blossoms turn the slopes pink and white.

Although Spain is practically self-sufficient in food production, its agricultural production is low by European standards. Farmers have sought to

expand production by using nitrogen as a fertilizer, which has strongly increased the nitrate content of river water, soil and produce. Research has shown that organic produce contains 93 percent less nitrate than non-organic produce.

Spain's organic agriculture has been officially regulated since 1989. Organic farms and food manufacturers can have their products and processes

certified according to the standards set by the CRAE (the national regulatory body for organic agriculture). To cope with the recent increase in applications (partly due to European Union subsidies), the inspection and certification process has been decentralized: each autonomous region now has its own certifying body and hallmark.

In 2000, 380,838 hectares of land were in use for organic agriculture, compared to only 4,235 ha in 1991. In this nine-year period, the number of organic farms increased from 436 to 14,040. These farms are found primarily in Catalonia, Andalucia, Extremadura, Aragón, Castillia-Léon and Valencia. Despite this growth, consumption of organic products is low in Spain and most of the produce is exported to Germany, Japan, France and Austria. However, Spain now has some 250 distribution companies, a few thousand organic food shops and several large supermarket chains and markets where organic products are sold. In 1992, the independent Spanish Organic Agriculture Association (SEAE) was established to promote and improve agricultural research and to provide education and advice about organic farming. The *Asociación Vida Sana* (Healthy Living Association) is also involved in promoting and developing organic farming.

Nature Conservation

Pesticides and nitrogen fertilizers are not the only threats to Spain's environment and natural habitat. Mass tourism along the coast and large-scale desertification in the interior are also taking their toll. A lack of (working) sewage and water treatment plants is seriously disrupting the ecosystem, particularly on the tourism-oriented Mediterranean coast. Fortunately, there is a growing awareness of the importance of nature conservation; more and more areas are being declared protected zones. Under Habitat 2000 scheme, many cities have endeavoured to make their surroundings more sustainable. Alcobendas (Madrid) now irrigates its public parks with waste water, which results in a 90 percent saving on the use of potable water. In Saguirren (Pamplona), 4,200 new homes with ecological climate control have been built, and 80 percent of Oviedo's inner city is now inaccessible by car. In Santa Coloma de Gramenet (Barcelona) - where the Besós river was an open sewer - wetlands have been created to naturally filter waste water and to enhance the landscape. Ten years ago, separating refuse was an unknown concept in Spain. Now, even the remotest villages have containers for the separate disposal of plastic bottles and aluminium cans.

Practical Information

Travelling to Spain

By bus: Eurolines buses go via Paris to Barcelona, San Sebastián, Vitoria, Asturia, La Coruña, Valencia, Alicante and other destinations. There are buses equipped to transport your bicycle to Girona and Cambrils (Catalonia), Irún (Euskadi), Santander (Cantabria), Logroño (La Rioja).

By train: Travel via Paris to Spain. The TGV (high speed train) will take you to Cerbères (on the French-Catalonian border), where you can board a Spanish train to Barcelona. From Paris, you can also take a Talgo train (luxury night train) to Barcelona or Madrid. For flights to Spain, please contact your travel agent.

Getting around in Spain

For information about train service, contact any RENFE railway station. The *Tarjeta turística* will buy you 8, 15 or 22 days of unlimited travel on all RENFE railway lines. Prices are available from the railway station.

Spain's high speed train is called AVE, and runs from Madrid to Sevilla.

You may take your bicycle on the train, but inquire in advance about conditions: bicycle transport depends on train type and time of day.

For bus travel, enquiries are best made at the point of departure. There are many different local and regional bus companies.

Sea travel: the Trasmediterránea shipping company operates a daily ferry service between the Iberian Peninsula and the Balearic Islands and Northern Africa, as well as a weekly ferry service to the Canary Islands. The main office is in Madrid, Tel: 91 431 07 00. Private companies also operate ferry lines from various coastal towns.

Food and drink

Meat and fish are the mainstays of the Spanish hot meal, but vegetarians are usually able to find a *tortilla* (potato omelette) or a vegetable dish on the restaurant menu. Spaniards have late dinners: midnight is no exception. Bars always serve the famous *tapas* (titbits).

Where possible, this guide includes information on the availability of vegetarian meals at places to stay in Spain. Remember that lunch is the main meal in Spain. Therefore places often charge more for lunch than dinner.

Playa Cala Fort near L'Ametlla de Mar, Tarragona, Catalunya

Catalonia and Aragón

Catalonia

Of all the autonomous regions of Spain, Catalonia (Catalunya) is the most independent, prosperous and politically headstrong. The region (whose capital is Barcelona) has its own language - Catalan - which is officially recognized and taught in schools. In some remote areas, Catalan is the only language spoken. Many places have two names: one in Spanish (Castillian) and another in Catalan, for example Lérida (Lleida), Gerona (Girona) and Figueras (Figueres).

Catalonia is not only home to the Costa Brava - Spain's number one tourist attraction - but also to the beautiful mountains of the Catalonian Pyrenees. In the breathtaking Aigüestortes national park, a perennial favourite among walkers, there are several ECEAT lodgings. There are also many 12th century Romanesque churches here, the most famous of which are in Taull. Hidden between the trees, their charming little bell towers are an integral part of the landscape. Most of the churches' original mediaeval frescoes have been removed from their walls or replaced with copies. The originals are now in the Museum of Catalan Art in Barcelona.

The national park, whose full name is Parc Nacional d'Aigüestortes i Estany de Sant Maurici, covers 10,230 ha. Its name comes from the twisted waters (aigüestortes) and the Sant Maurici lake (estany). This lake, magnificently located between the two mountain ranges of the Sierra de Encantats, is the starting point for many walking trails. These run north along smaller lakes and all the way to the peaks of the Agulles d'Amitges. The park has more than 150 lakes in all, dating back to the Ice Age. The deepest (50 m) and highest lake, Estany Negre, is in the southern part of the park.

The region has a rich biodiversity. In early summer, red and pink rhododendrons bloom in the lower valleys. Chamois make their home among the rocks and graze the mountain meadows; beavers and otters inhabit the lake shores. Eagles nest on the steep mountain slopes and capercaillie live in the forests. In summer, this region is excellent for walking, while in winter it offers ideal cross-country skiing. The park is accessible from Espot (eastern entrance) and Boí (western entrance); both entrances have visitors' centres.

Walking in the Catalonian Pyrenees.

La Garrotxa is a very different kind of park. This nature reserve is situated in the volcanic area around the town of Olot. In its entire 120 km², there are 26 smaller sanctuaries around the old volcanic craters. The area has a unique ecosystem. The crater hills vary in height from 200 to 1,100 m and stand in a green, hilly landscape dotted with mediaeval villages and castles. There is a visitors' centre with a botanical garden and a volcanic museum. Visitors may take walks, cycle, ride, or go swimming in the streams.

The Parc Natural Delta de l'Ebre, in the southern half of Tarragona province, is the largest and most important wetland area in Catalonia. It has fresh and salt water lagoons, river bank forests, sand dunes and rice paddies. The area's greatest treasure is its birds: no less than 300 species have been counted, and every year about 180,000 migratory birds spend the winter here. The park has an information centre where from excursions are organised.

Aragón

The once mighty kingdom of Aragón stretches from the Pyrenees in the north to the region of Castilla-La Mancha in the south. The towering peaks of the Pyrenees account for the extremes in the climate: long, cold winters alternate with hot summers. One of the features of the region is the river Ebro, with its fertile valleys. The Ebro has attracted peoples since ancient times, which is why Aragón has so many archaeological remains of Indo-European, Iberian and Roman cultures. Aragón also has many old towns with interesting historical sights. Everywhere you look you will see the Moorish Mudejar style, with its geometric decorations and detailed masonry: the best examples are found in Zaragoza and Teruel. The towers of Teruel are recognized by UNESCO as a World Historic Heritage site.

UNESCO has also granted protected status to some of the spectacular natural features and landscapes in the Pyrenean region: eroded rock formations, rushing waterfalls, enormous forests full of wildlife, deep gorges and ravines with tranquil or turbulent rivers, orchids and lagoons teeming with birds. Northern Aragón has the imposing Ordesa y Monte Perdido National Park, famous for its variety of vegetation and wildlife, such as the otter, chamois, eagle owl and various types of vulture. The park has many marked footpaths. Route descriptions are available at the park entrance, 9 km north of Torla.

1. El Molí

María Sanchís Pages
Mas Molí s/n,17469 Siurana d'Empordà,
Girona
Tel/fax: 972-52 51 39 Mob: 630-70 92 74
casaelmoli@teleline.es
www.turismerural.com/elmoli
Open all year Language: E, Ct, GB, F
€ pn B&B: 2p 49(ls)-54(hs); 3p 65(ls)-70(hs);
half board 2p 67(ls)-79(hs); 3p 96; dinner 9.02

Farm and surroundings

Mas El Molí is situated in the heart of the Alt
Empordà region, near the Costa Brava and Cap
Creus. The house is surrounded by 1 ha of
shade trees, medicinal plants and herbs, an orchard, a vegetable garden and a terrace. Many
birds love to visit the garden: nightingales,
golden orioles, goldfinches, blackbirds, bee-
eaters and others. Several different grains are
grown on the farm. There are around 100
Friesian cows, as well as chickens, rabbits and
other animals. This livestock and the vegetable
garden provide most of the food prepared in
the kitchen. María's exquisite dishes always
please the guests! At breakfast, you will enjoy
freshly baked cake and home-made cheese, yoghurt and jam. There are 6 large bedrooms,
each with a bathroom. Three of the bedrooms
have a balcony and one of them has an additional suite with a separate entrance, TV and
refrigerator.

María and José are happy to share their
knowledge of the region and its attractions.
During your stay you may want to visit the
nearby Dalí Museum in Figueres and Parque
Natural dels Aiguamolls de l'Empordà - wetlands of salt water lagoons and fresh water
lakes that provide refuge to thousands of
birds. Other worthwhile sights are the
monastery of Sant Pere de Rodes (15 km) and
Cadaqués. The beach at Sant Pere Pescador is
only 7 km away. There are many walking and
cycling routes in the vicinity of the farm (maps
are available). You can hire bicycles at El Molí.

How to get there

By car: The farm is 7 km SE of Figueres. On the
A7/E15 take exit 4 (S of Figueres) and follow directions to Vilamalla and Siurana. When you
reach the village, follow signs to 'El Molí'. **Public transport:** Train to Vilamalla (2 km from the
farm). Either take taxi, call to be picked up, or
walk 2 km towards Siurana and turn left just
before village. Walk another 500 m to El Molí.

2. Can Bosc

María Jesús Fernández Martínez de Ubago
& Richard Torrington
17745 Lladó, Girona
Tel: 972-54 70 92
canbosc2000@worldonline.es
www.canbosc.com
Open all year Language: Ct, E, GB, F
€ pn Self-catering: 120-162(ls) 138-186(hs);
Groups pp: 18

Farm and surroundings

Can Bosc is part of a tiny hamlet on the north
side of a secluded valley called Bac de La Sala,
some 4 km from the village of Lladó. Richard
and María Jesús moved here in 1997 with the
desire to establishing an ongoing permaculture
project. The buildings date back to 1775 and
were in quite a dilapidated state when they arrived. Over the years they carefully restored
them and started to work the land, following

ecological principles where possible, such as solar electricity use, solar heated water, ecological building materials, greywater recycling and plenty more. One of their projects entails a vintage-variety fruit orchard, including local species, which as they are not being produced commercially any more, are in danger of becoming extinct. A wide range of events, workshops and seminars is organised here, covering permaculture, belly-dancing, massage and mediation sessions.

The self-catering guesthouse includes five bedrooms, a fully equipped kitchen and a large sitting and dining room area with an open fireplace. In all it can sleep up to 9 people. Two attractive indoor rooms can be used by groups that wish to give their own courses. Mattresses, meditation cushions and a music system are available as well as extensive outdoor space.

Groups are catered for by arrangement. Vegetables, nuts, potatoes and olives are for sale.

Can Bosc is a perfect base for inland trips as well seaside activities, it is close to no less than three nature reserves (Cap de Creus, Aiguamolls d'Empordà and the volcanic Garrotxa), Roman sites and towns, cultural highlights like Girona, Olot and Figueres, with its Dalí museum.

How to get there:

20 km SW of Figueres. **By car:** N260 towards Olot, take exit Lladó (GIP5239). In Lladó (3.5 km to go) call to be picked up when you arrive. **Public transport:** train to Figueres. Pick-up by prior arrangement.

3. Rectoría de La Miana

Janine Westerlaken & Frans Engelhard
La Miana s/n, 17854 Sant Jaume de Llierca, Girona
Tel: 972-19 01 90
Open all year Language: E, Ct, GB, NL, F, D
Lodging pp: half board 33; full board 40; lunch 10; lunch-pack 7.25; VAT 7%; B!

Monastery and surroundings

Rectoría de La Miana consists of an old Romanesque presbytery and a Benedictine monastery. The site is graced by the ruins of a 9th century castle, La Miana, and a 12th centu-

ry Romanesque chapel. In the 1970s, Frans Engelhard discovered this idyllic place and began its restoration. After decades of decay, he breathed life into the property, which now exudes a mysterious, mediaeval atmosphere. There are various living rooms filled with antique furniture, galleries, a fireplace, and a large room to practice crafts and relaxation techniques. Various courses are on offer. The Rectoría has eight double bedrooms. The price for lodging includes a delicious breakfast and a good dinner. The house uses power generated by solar panels and water from its own source. It is set in a very peaceful area; the smell of the woods wafts up from the quiet, eternally green valley, which is home to wild boar. There is an excellent rock pool nearby where you can swim.

The surrounding mountains and the volcanic landscape of Olot are great for riding or walking (guided tours available). The mediaeval towns of Besalú and Santa Pau are worth a visit. The ancient city of Girona, Figueres (home of the Dali Museum), and the coast with its fishing villages and beaches are only an hour's drive away.

How to get there

By car: N of Girona, take the C66 and N260 towards Olot until Sant Jaume de Llierca. Drive into the village, then, from the Carretera Industria, drive 6 km on unpaved road until you find signs to Rectoría de La Miana. Follow signs until the end of the road. **Public transport:** Train to Girona, Figueres or Barcelona. Bus: from Barcelona take the Garrotxa Expres (TEISA), boarding at the corner of Pau Claris and Consel de Cent. You can also take a TEISA bus from Girona or Figueres. Get out at the Sant Jaume de Llierca stop. Guests are picked up only if strictly necessary.

4. Camping Masia Can Banal

Stendert Dekker
17855 Montagut, Girona
Tel: 972-68 76 81
(Tel/fax: 0570-61 45 61 in NL)
Open May-Sep 15th Language:
E, Ct, NL, GB, F, D,
€ pn Camping: tent 4-6.25; caravan &
camper van 6.25; adult 4; child 2; B!

Campsite and surroundings

This quiet, scenic campsite is 2 km from the village of Montagut, on the edge of La Garrotxa nature park. The 3 ha campsite is part of a 80 ha farm. The pitches are spread out over various meadows and terraces, or in hidden nooks behind the thicket. Fully furnished family tents are for hire. For caravans and campers, please seek advance permission (see telephone number below).

This environmentally friendly campsite has greywater treatment, separated rubbish disposal and uses solar energy (supplemented by a quiet generator in high season). In winter, cows and donkeys graze the campsite to keep the grass short and fertilize the grounds. In summer, they stay in the woods. The main building, an old Catalonian farmhouse, has an information centre, bar and restaurant, and a shop that sells organic produce and bread when available. Outside, there is a large shady terrace.

Children feel at home here. In summer, there are activities such as kite flying, clay modelling, theatre and survival trips. Advance booking is necessary, particularly in high season.

La Garrotxa's old volcanic landscape provides excellent walking. Nearby mediaeval villages such as Besalú and Santa Pau are worth visiting. Montagut has a public swimming pool, but a short walk will bring you to a natural basin in the river where swimming is great. Small children can use the wading pool on the campsite. The campsite has mountain bikes for hire and horses are for hire nearby.

How to get there

By car: From Figueres, take the N260 towards Olot. At km marker 74 turn right towards Montagut and Sadernes. In Montagut follow signs to Sadernes; after 300 m turn left at the sign to Can Banal. The campsite is another 1.5 km away. **Public transport:** Train to Figueres, then bus bound for Olot. Get out at Cruz de Montagut (5 km from campsite). Arrange a pick-up in advance via the Dutch booking office.

5. Camping Rural Els Alous

Armengol Gasull Queralt
Camí de Pineda 17856, Oix, Girona
Tel: 972-29 41 73
musica@bbs.grn.es
ww2.grn.es/musica/alous
Open Apr-Nov Language: GB, F
€ pn Camping: tent & caravan 3;
camper van 5.70; adult 3; child 2.70; VAT 7%

Campsite and surroundings

This small and beautiful campsite reserves most of its pitches for tents, but there are also a few caravan spots and a small number of caravans for hire. The campsite is in a lovely valley near the picturesque village of Oix. On the premises you will find a nice swimming pool, a plastic wading pool for young children and wide meadows for sunbathing or football. Rubbish is separated, and greywater is treated and used to irrigate a small apple and pear orchard. Vegetable and fruit waste is fed to the chickens and ducks.

In the spring and autumn, swimming is allowed in the natural stream that flows through the campsite. If you enjoy peace and quiet, try to avoid staying here during August: in this period many Spanish families spend their holidays at the campsite and their daily rhythm tends to differ considerably from that of most northern Europeans The owners strive to conserve nature and the environment. They occasionally organise nature walks and are happy to recommend interesting day trips. No less than two GR trails pass right through the valley.

The village of Oix, just 1 km away, has 2 restaurants and 2 small shops. The area is dotted with mediaeval villages. The campsite is 10 km from the Zona Volcánica de la Garrotxa, a nature reserve featuring a 350,000-year-old volcanic landscape.

How to get there

By car: Take N260 from Figueres to Olot/Besalú. One km before Castellfollit de la Roca (6 km before Olot), turn right towards Oix at a sign reading 'Camping Rural 10 km'. Just before Oix, turn left at 'Camping Masia' sign. Follow signs to campsite. **Public transport:** Train to Ripoll or Girona, then bus to Olot. Then bus to Oix, 1x weekly (at 1 km from campsite). Call in advance to be picked up in Olot.

6. Mas Pujou

Carmen Fernández & Pedro Escorihuela
Apartado 391 de Olot, 17800 Olot, Girona
Tel/fax: 972-19 50 76
pedcar@hotmail.com
Open all year Language: E
€ *pppn Groups: half board 17.44(ls)-19.25 (hs); full board 19.85(ls)-22.25(hs)*

Hostel and surroundings

This lovely old farmhouse is beautifully set in La Garrotxa volcanic park, surrounded by beech, oak and chestnut forests. Just a stone's throw from the famous La Fageda beech forest, this hostel is an ideal starting point for walking the park's many marked footpaths. Guided tours are possible. You can also discover the area on horseback from the nearby stables.

The old *masía* (farm) is sparsely furnished. The dormitories feature bunk beds, and the dining hall has long refectory-style tables. The common showers and toilets are very basic. The hostel can accommodate up to 50 guests. There is also one quadruple bedroom (with bunk beds). In spring and autumn, the hostel is used for groups of schoolchildren on ecology field trips. Other guests are always welcome, however. The cook prepares vegetarian meals on request, but not with organic ingredients. All electricity is generated by solar panels. Children can build their own solar-powered oven to bake pizzas.

Olot is an interesting town with museums of vulcanology and local fauna. There is a 57 km-long bicycle path that runs alongside the old Olot-Girona railway and passes through 12 villages and beautiful valleys.

How to get there

By car: From Figueres, N260 until exit 3 to Besalú and Olot. In Olot, C152 towards Les Preses. On roundabout before Nissan garage, turn left on to Calle Alzina. At smaller roundabout, turn right on to Calle Godua. Then take last street on the right: Calle Ginestola. This turns into unpaved Ctra de la Quintana, which passes by stables and Mas Gou. In the woods, turn left at sign reading: Mas Pujou 1200 m. **Public transport:** Train to Ripoll, Barcelona. The Teisa bus line provides regular service to Olot (2 km from hostel) from the large cities.

7. Mas Cabrafiga

Bakartxo & Juan Carlos Mahadeva
Sant Pere Despuig, 17813 La Vall de Bianya, Girona
Tel/fax: 972-29 12 68 Mob: 606-09 84 83
cabrafiga@mundofree.com
www.turismerural.com/cabrafiga.htm
Open all year Language: E, Ct, GB, I
€ *pw Self-catering: 480-720(ls) 540-825(hs); VAT 7%; B!*

Farm and surroundings

Mas Cabrafiga consists of an organic cattle feed farm and riding stables, situated on the southern flank of a wooded mountain. The farmhouse, which dates from 1362, has maintained its original character. Beautiful Spanish horses graze in the meadows around the farm. Mas Cabrafiga has three detached apartments which sleep four and six persons, each with a fireplace, kitchen, bathroom, garden and barbecue. The houses are on the sunny side of the mountain with a spectacular view of the valley. At weekends, the owners organise various excursions (for a small fee). They also sell organic fruit, honey, potatoes and vegetables. The farm is a fun place for kids with lots of safe space for outdoor play. The valley below has 16 Romanesque churches. The stream running alongside the Santa Llúcia del Puigmal church is excellent for swimming and fishing. There is an open air market every Monday in Olot. The town also has two museums. A festival for the patron saint is held every 7th and 8th of September. The nearby Zona Volcánica de la Garrotxa park is an interesting geological site. Many bird and mammal species live among the 30 volcanic craters and indigenous forests there.

How to get there

By car: From Figueres take the N260 to Olot. Immediately after Castellfollit de la Roca, take the GI522 towards Camprodon. Where this road joins the C26, turn right towards Camprodon. Just before Vall de Bianya, turn left and follow the signs to Cabrafiga. **Public transport:** Train to Girona, then the bus to Olot. Call in advance to be picked up.

8. Camping Mas la Bauma

Rob van Engelen & Wieske van Essen
Mas la Bauma s/n
17862 Vallfogona de Ripollés, Girona
Tel: 972-70 07 12
(020-688 14 91 / 06-513 441 87 in NL)
m.souwer@worldonline.nl
www.maslabauma.com
Open Apr-Oct Language: E, Ct, NL, GB, D
€ pn Camping (tents only): tent 4.50-6;
adult 4.50; child 2.70; tent to rent 320 pw;
breakfast 4.80; dinner 12.32

Farm and surroundings

Mas la Bauma is a small and environmentally friendly campsite on a farm. It is set in a valley in the foothills of the Pyrenees (elev. 1,000 m), surrounded by a mixed deciduous and evergreen forest with a wealth of flora and fauna.

The campsite was designed to have a minimum impact on the environment: you pitch your tent on terraced former potato fields. The campsite uses solar energy. Its name is derived from the large rock formation on the premises; *bauma* means hanging rock or shallow cave in Catalan. A nearby river has natural swimming holes. Guests can purchase various products on the campsite, including dried mushrooms, home-grown vegetables, herbs, eggs, paté and jam. The bar is open every day, and bar meals are available 3-4 times a week. Fully equipped, furnished 4-berth tents are for hire. This mountainous area is not accessible to caravans. Reservations in the Netherlands only Monday from 20.00-22.00 hours.

The surrounding area is a walker's paradise, from short strolls to mountain hikes on the highest ridges. There are plenty of clearly marked footpaths. Not far from the campsite

21

there are beaches and cultural sites to be found in cities such as Barcelona and Girona, and in smaller villages.

How to get there

By car: From Perpignan, France. Border crossing at La Jonquera, Salida 3 Figueres North. In Figueres, take the Besalú/Olot exit. In Olot, take the N260 towards Vallfogona. After km marker 102, turn left on to a narrow concrete road (downhill) at the sign for Mas la Bauma. Follow the signs. **Public transport:** Train to Ripoll or Figueres, bus from Ripoll (2x daily) or Olot (1x daily) to Vallfogona de Ripollés (2 km away).

9. Masia Serradell

Mercedes Palou Rifa & Martí Torras Palou
Masia Serradell s/n, 17530 Campdevànol, Girona. Tel: 972-73 09 50
Open all year Language: E, Ct, GB
€ pn Self-catering: 60(ls)-66(hs); extra bed 9.01; VAT 7%

Farm and surroundings

Masia Serradell is a stately old farm, which has been very well preserved. It includes buildings erected in 1300, and somewhere in this immense complex there is even an ancient chapel. The dining room, with its old furniture, breathes an atmosphere of earlier times. There are double bedrooms in authentic style and apartments with their own fireplaces. The view is beautiful on all sides. This farm is run with all the hospitality that Catalunya is famous for. You eat your meals with your hosts at the family table. The hosts make a living from timber and extensive cattle raising. They plan to start

using solar energy to generate electricity and heat water. The farm is situated near La Sierra de Gombrènx. You can visit nearby villages and take walks in the woods.

How to get there

By car: From Ripoll take the N152 towards Ribes de Fresner. In Campdevànol turn left immediately on to the GI401 towards Gombrèn. At km marker 5 there is a sign to Masia Serradell. Take this exit and after 1 km of unpaved road you will reach Masia Serradell. **Public transport:** Train from Barcelona to Campdevànol, then take a taxi (4 km). Or bus from Ripoll to Campdevànol, get off at the L'Empalme bus stop. Walk the remaining 1 km, take a taxi or call in advance to be picked up.

10. Camping Molí Serradell

Eduard Torras Manso
Apartado 17, Sant Llorenç, 17530 Campdevànol, Girona. Tel/fax: 972-73 09 27
calrei@teleline.es
Open all year Language: E, Ct, GB, F
€ pn Camping: tent, caravan, camper van 8.50; adult 4.50; tent to rent 6; lodge to rent 61; VAT 7%

Water mill and surroundings

This campsite (established in 1971) is beautifully set on the bank of a small river which once powered the water mill. The old mill now houses a bar and a restaurant, which serves good, traditional meals. Bread is baked in an old wood-burning oven.

The owners have many sheep and cows that graze in the mountains in the summer. These animals and the pigs, chickens and rabbits are the basis for the regional dishes served in the restaurant.

The campsite has 55 spaces (of approx. 60 m2) which house mainly caravans in the winter and tents in summer. Hikers travelling with smaller tents can set up camp in a separate clearing in the woods. There is a playground and large swimming pool. Swimming is also possible in the river. For those willing to take a 1-hour walk, the river also has waterfalls, and there are places to swim there. The campsite is beautifully laid out with paved roads and lots of wood. The sanitary facilities are well maintained and clean. The campsite also has a number of wood cabins for hire. This area in the foothills of the Pyrenees is an excellent place to walk.

How to get there

By car: From Ripoll take the N152 towards Ribes de Fresner. In Campdevànol turn left immediately on to the GI401 towards Gombrèn. After km marker 4, a road off to the left leads you to the campsite. **Public transport:** Train from Barcelona to Campdevànol, then taxi (5 km). Or bus from Ripoll to Campdevànol, get off at L'Empalme bus stop. Walk remaining 1 km, take taxi or call in advance to be picked up.

11. Camping Masia Can Fosses

Lluis Rodriguez i Pont
Can Fosses s/n, 17535 Planoles
Girona. Tel/fax: 972-73 60 65
Mob: 629-78 06 79
Open Apr-Oct Language: E, Ct
€ *pn Camping: tent 2.84-3.31; caravan 3.31; camper van 6-6.62; adult 3.31; child 2.48; tent to rent 6*

Campsite and surroundings

This campsite (elev. 1,200 m) is on the premises of Masia Can Fosses, a traditional mountain farm that has been in the same family for over 500 years. Around 100 Pyrenean cows and calves graze on the slopes of the property, which is surrounded by 100 ha of red and black pines. The terraced campsite is on south flank of the mountain and provides plenty of sunshine, tranquillity and panoramic views. There

are about 25 pitches for tents and caravans. There are also 3 tents for hire. The campsite has a common room with an open fireplace and a cafeteria/bar. Lluis has lived here all his life and knows the region like no one else. He will take you on guided tours to the caves known as 'Las Grutas de las Encantadas', on expeditions up Mount Puigmal (2,913 m), and on walks through the protected woods where there are many centenarian pine trees. Children may help giving salt to the cows.

A playground, swimming pool, football field, tennis court, shop and restaurant are all to be found in Planoles (1 km away). There are plenty of walking and cycling routes (the campsite is on the GR-11) and information is available at Can Fosses. A small rack locomotive will take you through the breathtaking valley of the river Núria up to the isolated sanctuary of Nostra Senyora de Núria. Ski slopes are located 7 and 12 km away.

How to get there

By car: From Barcelona take the C17 and N152 towards Puigcerdà to Planoles. At km marker 127 take the small road to the campsite (0.5 km). **Public transport:** On the Barcelona - Puigcerdà line, train to Planoles (about 1.5 km from the farm). Then taxi, call to be picked up or walk about 2 km. Or bus to Planoles; get off at the bus stop on the N152 (1 km from farm).

Spain

12. Cal Pastor

*Josefina Soy Sala & Ramón Gassó Germe
Calle Iglésia 1, 17536 Fornells de la
Muntanya (Toses), Girona
Tel/fax: 972-73 61 63
ramongasso@logiccontrol.es
Open all year Language: E, Ct, F, GB
€ pn Lodging: 2p 36.06; 3p 39.07;
Self-catering per unit: 78.13; breakfast 4.51;
dinner 10.52; VAT 7%; B!*

Farm and surroundings

The 200 year old Cal Pastor farm was restored
in 1993. Surrounded by mountains, streams
and meadows, it is one of the few buildings in
a tiny mountain village that is home to only
three families. The farm owners make a living
from extensive cattle raising: for generations,
the men of the family have been herding cattle. Through the years they have kept many of
the old tools of their trade: their cups, clothing,
equipment and bells. These objects are on dis-
play in a small museum adjoining the house.
The house itself is comfortable and nicely decorated with lots of woodwork. There are double, triple and quadruple guest bedrooms, and
2 apartments big enough for 5 guests each.
Breakfast and dinner are available on request.
Guests are welcome to participate in the farm
activities. Situated in the middle of the Pyrenees, there are plenty of outdoor activities in
summer. In the winter, there is good skiing in
the vicinity. La Molina ski area is 12 km away.

How to get there

By car: From Ripoll, take the N152 N towards
Planoles and Puigcerdà. At km marker 133.5,
turn left down a descending road. After about
3 km, you will reach Fornells de la Muntanya. In
the village, turn left and then immediately to
the right. The large house on the right is Cal
Pastor. **Public transport:** Train from Barcelona
towards Puigcerdà, get out at Toses station.
From there, take a taxi, walk (3 km) or call in
advance to be picked up.

13. Mas Can Sala

*Cecile Kraetzer & Camillo Sturm
Can Sala, Granollers de Rocacorba
17153 Sant Martí de Llémena,
Girona. Tel: 972-44 31 62 Mob: 646-96 78 11
Fax: 972-44 32 14 infocel@luzdevida.com
Open May-Oct Language: NL, D, GB, F, E, Ct
€ pn Camping: adult 7(ls)-10(hs); Lodging pp:
25(ls)-35(hs); breakfast 4.80;
lunch & dinner 7.30*

Farm and surroundings

Twenty five years ago Cecile and Camilo helped
to found *De Kleine Aarde,* a Dutch centre for
ecological awareness and sustainable farming.
They later moved with their five children to
Spain, where they set up Luz de Vida, now
Spain's biggest distribution centre for organic
food. Their latest project is Llemenà, an ecology centre situated on a beautiful, 175 ha property with a large, centuries-old manor house.
The centre has a shop which sells Luz de Vida
products and an organic restaurant. Fifteen ha
of land is dedicated to organic agriculture and
permaculture. Chickens, geese and rabbits
roam freely on the premises. The project is still
under development; the owners are experi-
menting with various ecological applications
such as sustainable building materials and solar
energy. They plan to offer courses in organic
farming. Five beautiful guest rooms are al-

24

ready complete. The showers and toilets are for common use. There is a small area available for those who wish to pitch a tent. Swimming is permitted in the *embalse* (reservoir). The rooms are open to groups year round.

There are excellent walks on the grounds themselves, and many marked routes in the surrounding area. For cultural activities, Girona is nearby.

How to get there

By car: From Girona take the GI531 to Sant Martí de Llémena. Two km the village (just before km marker 21), turn right towards Granollers de Rocacorba. The entrance to the property is approx. 1.5 km down this road. **Public transport:** Train to Girona, then the bus to Sant Aniol de Finestres (Roca bus company; 3x daily on weekdays, 2x daily on weekends). Ask the driver to let you off at CEL.

14. Masia Les Planes - Paisatge i Aventura

Joan Massagué Planas
Masia Les Planes s/n, 08263 Sant Mateu de Bages, Barcelona
Tel: 938-36 00 60 Fax: 938-36 04 59
carros@mx3.redestb.es
www.minorisa.es/aventura
Open all year Language: E, Ct, F
€ Horse-drawn wagon: 200-225 2 days;
extra day 80-85; breakfast 4.80;
lunch & dinner 10.85; B!

From farm to farm by horse-drawn wagon

This is not a place to stay, but a trip you can take by horse-drawn wagon through a rustic, attractive nature area. The horse and wagon can be hired from 2 days up to 2 weeks. The first leg of the journey is always with a guide, and from there on the horses know the way. Every day you cover 15 to 20 km and arrive at another farm where your horse is kept in the stable. Your lodgings are in the wagon itself, which is fully equipped for 4 people, with a table, beds, a tent lantern, a heater and a water tank with a wash basin (bring your own sleeping bag). You can cook your own dinner if you book a wagon with cooking facilities, or eat at the farms you visit: breakfast, lunch and dinner are catered for if booked in advance. If you wish, you may also stay longer in one location. You will pass by many sights along the way, including a Romanesque chapel, beautiful woods and castle ruins. But above all, the landscape itself is simply breathtaking. Before you leave, the wagon is completely equipped and you are given full instructions. And then, off you go! You will also be given information about natural springs, Romanesque churches, lodgings, history, gastronomy, flora and fauna, and local events. You can also go on expedition on a donkey.

How to get there

By car: Drive from Barcelona via Manresa to Callus (C55). After 5.5 km head for Sant Mateu de Bages (BV3003). On the left side there will be a sign reading Paisatge i Aventura. Follow sign to Masia les Planes farm. **Public transport:** Train to Manresa, then the bus to Callus. From there take a taxi (6 km).

15. La Torre de Guialmons

Lluis Marcet Pacios
Guialmons, 43428 Les Piles, Tarragona
Tel: 977-88 05 58
maslatorre@navegalia.com
Open all year Language: E, Ct, GB, F
€ pppn Lodging: 18.60; full board 39-43.20;
Self-catering: 18.60; breakfast 3.60;
dinner 9.60

Farm and surroundings

Mas la Torre de Guialmons is a walled farmhouse built in 1905. Lluis and his family purchased the house in 1995. They renovated the

25

Spain

building, taking care to preserve the traditional architectural style and furniture. Lluis's main goal is to become completely self-sufficient. He has created a small ecosystem within the walls, with a large garden, an organic vegetable garden, chickens, many birds, medicinal plants and herbs, and exotic stones and minerals. Lluis intends to install solar panels for electricity and hot water. The stables have been converted into a self-catering apartment with 4 bedrooms with en suite bathrooms, a kitchen and a common sitting/dining room with an open fireplace. Home-grown and local organic products are sold. Breakfast and dinner are served. In the evenings, Lluis is happy to share his expertise on organic farming and treat you to the sweet melodies of his guitar.

The Conca de Barberà region is great for walks (GR-176 and 7b), bike rides and excursions. At the farm you will find information on the Wine Route (Ruta dels Cellers); the Cistercian Route with the monasteries of Poblet, Santes Creus and Vallbona de les Monges; the Mediaeval route to Guimerà, Conesa and Castell de Biure; the Castle route along the river Gaià and the Mill route along river Corb with its 51 mills.

How to get there

By car: Take the N-II from Barcelona towards Lleida. At Igualada, take the C37 S towards Valls. After approx. 18 km turn left to Sta. Coloma de Queralt (B220). From here take the C241 towards Montblanc and after 2 km you will reach Guialmons. **Public transport:** From Barcelona take a train or bus to Igualada, then a bus to Sta. Coloma de Queralt. From there, take a taxi or call to be picked up.

16. Mas de Caret

Teresa Rosell & Carles Llurba
Ctra. La Riba - Farena km 11
43459 Montblanc, Tarragona
Tel: 977-26 40 03 Mob: 689-26 47 95
www.pradesmontsant.com
Open all year; camping Mar-Dec
Language: E, Ct, GB, F
€ *pp: half board 30(ls)-33(hs); full board 39(ls)-43.02(hs); Self-catering: 120.20(ls)-150.20(hs) weekend; 210.30(ls)-252.40(hs) pw; lunch & dinner 9.10; B!*

Farm and surroundings

Mas de Caret farmhouse is in the hamlet of La Bartra (elev. 675 m), in the heart of the Prades mountains. It was built in 1786 and was recently restored by owners Teresa and Carles according to ecological principles. The farm gets its electricity from wind and solar power, and its greywater is naturally purified. Your hosts Teresa and Carles will be happy to explain how these facilities work.

The house is in the middle of the forest. Guests who respect the environment may camp out here (room for 4-6 tents). A caravan and 2 tents are for hire.

The farmhouse has several guest rooms and 1 self-catering apartment (for 6-8 guests). There are 3 quadruple bedrooms with *en suite* bathrooms, 1 double bedroom (extra cot available) and a patio. There is a large dining room/lounge with an open fireplace, a swimming pool and a playground. No smoking is permitted in the common rooms. Chickens, ducks, geese and a pony live on the farm. The owners serve delicious meals (vegetarian on request) prepared with home-grown organic produce. They also sell fruit, nuts, olives, honey and herbs. Teresa and Carles love to share their knowledge of organic farming, ecological building and sustainable energy. Guests staying at least a week can help out on the farm for room and board.

Besides enjoying the quiet rural setting, you can also opt for an active holiday. The Montanyes de Prades offer good walking, riding (8 km away) and mountaineering (4 km away). Sights worth seeing include prehistoric cave drawings (3 km), the mediaeval village of Montblanc, and the Poblet monastery on the Ruta del Císter (Cistercian route). The ancient forest surrounding the monastery is home to many interesting trees, including a willow unique to southern Tarragona.

How to get there

By car: Mas de Caret is 45 km NW of Tarragona. From Reus, take the C14 to Montblanc. After approx. 26 km, turn left to La Riba and Farena (TV7044). At km marker 11, before you reach Farena, take the road to La Bartra. **Public transport:** Bus or train to La Riba (12 km away). Then taxi or call in advance to be picked up.

17. Mas de Mingall

Ansje Abelsma & Klaas de Poel
Apdo de correos 49, 43519 El Perelló, Tarragona
Mob: 659-63 46 84 (020-637 46 39 in NL)
Open all year Language: E, GB, F, D, NL
Self-catering: €1200 pw

House and surroundings

Mas de Mingall (elev. 175 m) is a large 19th century farmhouse standing in a valley with 14 ha of olive, almond and carob trees interspersed with pine woods. It remains green all year round in the valley, which is home to wild boar and many bird species, including golden orioles, nightingales and owls. Wild herbs and plants fill the garden with a lovely aroma. There are quiet, starry nights in this wonderful location in the great outdoors. The house has 5 bedrooms (one has a balcony and another has access to a roof garden) and 2 bathrooms. There is an open fireplace and a piano in the large sitting room. The large rustic kitchen has doors that open out on to a shady patio. The

garden has enough space for 3 tents, and there is a swimming pool surrounded by olive trees. Clear spring water is pumped up from 200 m underground and solar panels supply electricity. If enough guests participate, your host will arrange courses in mosaic, stained glass and pottery.

There are several walking and cycling paths nearby (guests may borrow up to 5 mountain bikes). From Mas de Mingall, you can trek across the mountains to mediaeval villages such as Tivissa and Miravet. The authentic fishing village of L'Ametlla de Mar and many quiet beaches are just a 15 minute drive away. Twenty km S of the farm is the Parc Natural del Delta de l'Ebre, an important wildlife and bird sanctuary.

How to get there

By car: 62 km SW of Tarragona. On the A7/E15, take exit 39 (L'Ametlla) and turn left towards Perelló. When the road curves left towards Perelló, go straight, on an old, bumpy asphalt road (TV3025). Continue for about 4 km, until T-junction, and turn right. Drive for 200 m, then turn left at large house called Mas Plate. Take this road for 4 km until sign for Mas de Rabassó. Turn left, and after 300 m, at a white post turn left to Mas de Mingall. **Public transport:** Train to L'Ametlla de Mar then taxi (12 km). Or call in advance to be picked up.

18. Casa Elisa & Can Torres

Juan José Bel i Roig
C/ Catalunya 14, 43878 Masdenverge,
Tarragona
Tel: 977-71 91 89 Mob: 616-29 10 50
agroturisme@casaelisa.com
www.casaelisa.com
Open all year Language: E, Ct, GB, F
€ *pn Lodging: 2p 32; Self-catering: 300.20-*
400.60 weekend; breakfast 4; lunch 12;
dinner 10

Houses and surroundings

Can Torres and Casa Elisa are 2 large self-catering houses (max. 15 guests each) centrally located in the quaint old village of Masdenverge. Both houses are in the typical architectural style of the region. Each house has 2 double bedrooms with *en suite* bathrooms, 5 bedrooms that share 2 bathrooms, a kitchen and 2 living rooms. Outside, there is a patio with a barbecue area. Each house is ideal for 2 families or 1 large group. The owners, who also run a small farm, sell home-made preserves and serve their guests home-cooked traditional meals made with their own organic produce. They also promote environmental awareness and sustainable rural tourism. Juan organises guided excursions and teaches wood-carving and how to make preserves.

Masdenverge is in El Montsià, a region bordering on the Delta de l'Ebre nature reserve. The tiny village, with its narrow streets, still has a town crier who broadcasts important events through loudspeakers to the entire village. Amposta is 5 km away, on the banks of the river Ebro. The town is surrounded by rice paddies and has an interesting museum displaying local archaeological finds. It is also the gateway to the Ebro estuary nature reserve (Parc Natural Delta de l'Ebre).

How to get there

By car: On the A7/E15, take exit 41 to Amposta. From Amposta, to Uriola and Masdenverge (T344). **Public transport:** Santa Bàrbara railway station is 2 km from Masdenverge. Take a taxi or call (in advance) to be picked up.

19. Venta de San Juan

Clotilde Pascual
Venta de San Juan s/n, 43786
Batea, Tarragona
Mob: 649-64 47 24
www.agronet.org/agroturisme/sanjuan
Open all year Language: E, Ct, GB
Lodging € *pp: 18; B&B 21; half board 33;*
extra bed 9; Self-catering € *pn: 120 weekend;*
300 pw; add. cleaning; VAT 7%

Farm and surroundings

The impressive Venta de San Juan manor was built in 1898 by a Cuban who stayed on in Spain after the Spanish colony of Cuba gained independence. The estate has 125 ha of land; 40 ha are used to grow organic almonds, olives and wine grapes. Fifty sheep and goats graze amid the trees. On the first floor of the house there are 3 large rooms for 5 guests each and shared bathrooms. The rooms have a splendid view of the surrounding countryside. Casita de Vendimiadores ('Grape pickers' cottage') is a self-catering apartment with 2 bedrooms and 1 bathroom, a sitting/dining room with an open fireplace and a fully equipped kitchen. Breakfast and dinner are served in the manor house dining room. There is also a lounge that opens out on to the garden. The house is lit by candles and torches in the evening, and a generator is available if electricity is needed.

You can take walks and bike rides (bicycles are for hire) on the estate. There are pine forests that smell of rosemary and thyme, and the colours of the vineyards and orchards change beautifully with the seasons. The pretty villages of Batea (with an open-air swimming pool) and Fabara are 8 km away. Both can be reached on foot. Sights to see include the ruins in Pinyeres and Horta de Sant Joan, a 2nd century Roman mausoleum in Fabara and a castle of the Knights Templars in Miravet.

How to get there

By car: From Reus (near Tarragona), take the N420 W. After 76 km (6 km past Gandesa), turn right on to the C221 to Batea. From Batea, take the T723 to Nonaspe. At km marker 7.3, turn left on to an unpaved road (before the sign reading 'Provincia de Zaragoza'). Continue on this road for 2 km (keeping left) until you reach Venta de San Juan. **Public transport:** The closest railway station is in Nonaspe (12 km), the closest bus station is in Batea (8 km). From either station, take a taxi to the manor house.

20. Mas Lluerna

Ricard Guiu Peiró & Isel
Partida Pedregals s/n, 25617 La Sentiu de Sió, Lleida.
Tel/fax: 973-29 20 12
Mob: 652-27 35 54 ridolet@terra.es
Open all year Language: E, Ct, F, GB
€ *pn Camping: tent 4.80-6; caravan & camper van 7; adult & child 4; Lodging: 2p 36.06; 4p 12.02 pp; Self-catering: 6p 90; dinner 10; B!*

Farm and surroundings

This organic farm nestles among holly oaks and looks out over a gorgeous valley. The farm was built entirely from environmentally friendly materials. Owners Dolores and Ricard grow fruit and vegetables and keep bees. The farm doubles as a nature and environment education centre, offering courses and activities pertaining to organic agriculture, alternative energy sources and herbal medicine.

There are 2 double bedrooms, 1 of which adjoins a private living room. Guests share a bathroom, recreation room and terrace. There are also a six person apartment and a quadruple bedroom with bunk beds, which can be shared with other guests (bring your own sleeping bag). Use of bio-degradable cleaning and skin care products is free of charge. The 2 showers are outside. The farm has a collection of some 100 ecology games in 5 languages, as well as a library and a slide projector. There are many beautiful, quiet spots on the farm where guests can pitch their tent. The water reservoir across the road is ideal for swimming. Guests are requested to conserve electricity, which is supplied by solar panels, and to smoke outside. Meals are provided on request. A large variety of organic farm produce is for sale and the kitchen can be used for a fee.

The Sierra del Montsec offers good mountain climbing, while the Rio Segre allows you to canoe and fish. The area is also good for walking and discovering medicinal plants and a wide variety of bird species (there is a lammergeyer sanctuary). In the vicinity there are ruins of mediaeval settlements and villages with Moorish, Roman and Gothic architecture.

How to get there

By car: Situated at 10 km NE from Balaguer. From Balaguer take the C13 and C26 towards La Seu d'Urgell. After the bus stop at km marker 32, turn right after km marker 33 on to an unpaved road at the sign to Finca Bensa. Follow the signs to Mas Lluerna (3 km to go). **Public transport:** Train from Lleida to Balaguer (2x daily). Then take the bus (2x daily) to La Sentiu de Sió. Get off at bus stop at km marker 32 on the C26. Then walk following indications above.

21. Casa Pete y Lou

Lou Beaumont & Peter Dale
San Salvador de Tolo s/n, 25638 Tremp,
Lleida. Tel/fax: 973-25 23 09
Open all year Language: GB, E
€ pn B&B: 1p 16.75; 2p 33.50; lunch 6;
dinner 9-15; lunch-pack 4.5; B!

House and surroundings
In the South East corner of the Tremp Valley, Pete and Lou run their traditional farmhouse, which is perched on a forested hillside. The house is surrounded by and overlooks wheat, corn and sunflower fields, interspersed with ancient almond trees and olive groves.

Rooms for B&B are on the first floor, comprising two doubles and one twin-bedded room and a private sitting area with coffee and tea making facilities. Shower is shared with the other guests. Meals are served in the kitchen , which is the focal point of the house. Bread is home-made, and organic vegetables and fruits come straight from their own kitchen garden. As cooking is Lou's passion, you may want join for evening meals. The local vineyards produce delicious wines.

Casa Pete y Lou sits at 1,000 m altitude, just north of the wonderful Montsec escarpment -

where Mediterranean flora and fauna meet their alpine equivalents. The slopes are cut by the dramatic gorges of Tarradets and Vilanova de Meiá; the latter is at a 30 minute walk distance. The area is renowned for its romanesque architecture, diverse geology and is a site of dinosaur footprints and fossilised egg remains (Isona). The best time to visit depends on your preferences, as all seasons have an incredible variety to offer: spring brings immense numbers and varieties of wild flowers (including orchids and ophrys) and birds, such as the hoopoe and bee-eater, which you can identify with the help of the well-stocked library at the house. High mountain ranges are found at the Parc Nacional d'Aiguestortes and Val Fosca, where Pete loves to walk. As an experienced mountaineer he is available to take you on escorted walks by arrangement.

How to get there
From Barcelona Airport follow signs for Barcelona, turn right on to N11, A2 Tarrega-Lleida, continue until Tarrega. Turn off on to C14 to Agramunt and Artesa de Segre. Here turn left on to L512 direction Tremp and Isona. Continue uphill for about 30 mins and turn left at defunct petrol station Cap de Serra on to L911 towards Sant Salvador de Toló. After 6 km (immediately behind the 2nd set of crash barriers on bridge) turn off at track signposted Casa Pete y Lou. Continue till it forks, take right fork and continue till track splits in three and you see another sign to Casa Pete y Lou. Park here.

22. Cal Sodhi

Rosa Fàbrega & Josep Fàbrega Buchaca
Argestugues, 25795 Noves de Segre
(Les Valls D'Aguilar), Lleida
Tel/fax: 973-29 83 07 Mob: 609-67 52 38
rfabrega@mundivia.es
www.euroconnect.co.uk
Open all year Language: E, Ct, GB, F
Self-catering: €300 weekend; €660 pw; B!

Farm and surroundings
Cal Sodhi is a converted 18th century stone and wood hay barn (elev. 1,000 m) with a breathtaking view of the Pyrenees. It is the only inhabited house in Argestugues, a hamlet with 3

other houses which were deserted in the 1960s. There are chickens in the farmyard, and goats and sheep grazing the surrounding fields. Children may help herd the animals and feed the hens. Cheese and vegetables are for sale. The large garden offers splendid views, plenty of sun and pure mountain air. The cosy house has 4 double rooms, a dining room and lounge with an open fireplace and a fully equipped kitchen. The lodgings are available for groups of 4, 8 or 10 guests. For electricity, the house uses solar and wind energy in summer, and a water turbine in the winter. The owner can tell you a great deal about the mountains, the local flora and fauna, local history and legends. He organises field trips to the mountain peaks in search of wildlife, and at the end of the day he will guide you to one of the small village restaurants for a traditional Pyrenean meal.

L'Alt Urgell offers a combination of nature, culture and folklore. The area is great for walking, cycling and riding. Sights include Roman buildings and prehistoric burial sites. There are numerous village festivals to enjoy.

How to get there

By car: Argestugues is just S of Andorra, 17 km SW of Seu de Urgell. On the C14 from Lleida to Seu de Urgell, at km marker 172, turn off on to the road to Noves de Segre. Once there, ask for directions to Argestugues. **Public transport:** Take a train to Puigcerdà. From there, it is another 50 km by bus to Seu de Urgell. From there, take a taxi or call in advance to be picked up.

23. Camping Masia Bordes de Graus

María Angeles Biarnès Biarnès & Salvador Tomas Bosch
Crta Pleta del Prat km 5, 25577 Tavascán, Lleida. Tel: 973-62 32 46
Open Apr-Nov Language: E, Ct, GB, F
Camping, lodging & meals: prices on request

Farm and surroundings

The small but pretty campsite is situated in the meadows next to a little river. A beautifully restored old farm now houses a bar, restaurant, washrooms for the guests, and the owners' living quarters. All the buildings are built of stone and wood. The hay barn has been converted into nicely decorated bedrooms and a large dormitory. The campsite is mainly used for tents, and seldom for campers or caravans. The little river offers swimming, there are swings for the children and fields to play football and other games. The owners make a living from raising animals; sheep, goats and horses graze the mountains in the summer. Meat dishes served in the restaurant come from the farm's own herds and vegetables from the owner's grandparents' organic garden in the neighbouring village. Grocery shopping is done for you; all you have to do is order what you need. The friendly owners know the surroundings very well and are happy to help you plan a walk (maps are available). The location of this campsite, in the heart of the Pyrenees at 1,360 m elevation, makes this a great starting point for hiking through the unspoilt mountainous nature - plenty of eagles, and chamois - and through almost deserted hamlets. You can take advantage of a large selection of guided activities such as rafting, mountain climbing, skiing and riding.

How to get there

By car: From Sort, take the C13 to Llavorsí. There, take the L504 towards Ribera de Cardós. Stay on this road until you reach Tavascán. Just past this village is a sign pointing to the campsite. Turn left on to the asphalt road, which becomes a dirt road (still passable). From there it is another 1.3 km to the campsite, in the direction of the cross country ski trails. Follow the signs. **Public transport:** Train from Lleida to La Pobla de Segur, then the bus to Llavorsí. From there, take a taxi or call in advance to be picked up.

24. Casa Tonya

María Carmen Picolo Cuero
Calle Única s/n, 25588 Unarre, Lleida
Tel: 973-62 60 39
Open Jul-Oct Language: E, F, Ct
Lodging & meals: prices on request

House and surroundings

Casa Tonya is a typical village home restored by the owner in 1990. Antique and locally manufactured furniture create a very tasteful interior. The owner also has a restaurant, beautifully decorated in traditional style, where she serves time-honoured dishes using home-grown produce. She also makes her own sausage, wine and desserts, and she is happy to let guests take a peek in the kitchen. Unarre is a small mountain village in the heart of the Pyrenees whose inhabitants still follow their traditional ways. The Pyrenees offer many wonderful walks through unspoilt nature. The village is close to the Aigüestortes National Park. Spain's biggest ski resort, Baquiera Beret, is just 40 km away.

How to get there

By car: From Sort, take the C13 N to Llavorsí, then continue to Esterri d'Àneu. Just before reaching this town, turn right at the road sign pointing to Unarre (another 4 km). In the village, Casa Tonya is the white house up on the left. **Public transport:** Train from Lérida (Lleida) to La Pobla de Segur. Then take the bus to Esterri d'Àneu. Walk the last 4 km or take a taxi.

25. Camping Solau

Ctra San Maurici s/n, 25597 Espot, Lleida
Tel: 973-62 40 68
Open all year; camping Apr-Nov
Language: E, F
€ pn Camping: tent & caravan 3.75; camper van 7.50; adult 3.75; child 3.20; Lodging 16 pp
Self-catering: 72; VAT 7%

Farm and surroundings

This small farm/campsite is on the edge of the village of Espot, just 3 km from the splendid Parc Nacional de Aigües-Tortes i Estany de Sant Maurici. This park is famous for its lakes. There are marked footpaths and mountain cabins where you can spend the night. Tents, caravans and campers are allowed on the campsite. There are also bedrooms and small apartments for hire. The owners raise a herd of about 25 horses for meat production. They graze freely in the mountains from June until November, when they are rounded up to return to the stables.

The farm also has chickens and rabbits. There is good fishing in the river that runs alongside the campsite. Espot is a mountain village geared towards (winter) sports tourism. It has a tourist information centre, a few restaurants and shops, and is the starting point for trips to the national park. In winter, there are good ski slopes 2 km away.

How to get there

By car: Drive from Sort to Llavorsí. There, take the C147 towards Esterri d'Àneu and Espot. After about 12 km, turn left towards Espot. Once in the village, cross the bridge and turn right. After 500 m cross the old Romanesque bridge and turn right again, then take the first left. The campsite is 200 m down the road, on the right-hand side. **Public transport:** Train from Lerida (Lleida) to La Pobla de Segur. Then take the bus to the last stop before Esterri d'Àneu, across from the exit to the Aigüestortes National Park. From there, it is a 7 km walk.

26. Casa Teixidó

Josep Francés Palacín & Inés Tudel Perna
Calle Única 7, 22583; Molins de Betesa, Huesca. Tel: 974-34 71 43
Open all year Language: E, F
€ *pn Lodging: 2p 18-24; 3p 30; 4p 36.05; extra bed 6; use of kitchen 3; breakfast 4.20; dinner 8.40*

Farm and surroundings

Casa Teixidó (elev. 900 m) is an 18th century farm restored to its former glory in 1994. Its traditional decor and authentic furniture breathe the original atmosphere of rural Spain. The old kitchen has been preserved, and the new kitchen still has its original fireplace. The house is next to a stream in the unspoilt mountains of the High Pyrenees. Some farm animals roam the yard (sheep, rabbits and chickens). There are 2 double bedrooms and 1 quadruple that share a bathroom. There are also 2 triple bedrooms with adjacent bath, and a living room with kitchenette. You may opt for half board or cook your own meals (small fee for use of kitchen). Your hostess prepares enjoyable dishes using fresh produce from her own vegetable garden. Vegetarian dishes are served on request. Guests may also hire the entire accommodation. Children can play safely along the stream and around the house.

With its central location, Casa Teixidó is an ideal point from which to set out on various

day trips. You can walk in the Sierra de Sis, or walk to the Barranco de Viu gorge and swim in the Pantà d'Escales. For winter sports, there are ski slopes just 40 km away.

How to get there

By car: On the N230, turn off at km markers 112-113 (in a bend, 14 km S of Pont de Suert). Follow this road for approx. 300 m until you reach the sign for 'Betesa - Santoréns'. Take this road and after 2 km take the road to Betesa. Follow the wooden signs to TR (Turismo Rural) and you will find Casa Teixidó. **Public transport:** information available on request.

27. El Rancho de Boca la Roca

Hilde van Cauteren & David Samper Enjuanes
Boca la Roca s/n, 22580 Benabarre, Huesca
Tel: 974-54 35 65 Mob: 609-35 98 13
equipirineo@pirineosur.com
www.equipirineo.com
Open all year Language: E, GB, NL
€ *pn Camping (tents only): tent 2.70; adult 0.90; child 0.40; Lodging: 1p 30.05; 2p 36.06; 3p 45.07; 4p 54.09; breakfast 3.01; lunch & dinner 9.02*

Farm and surroundings

El Rancho de Boca la Roca is situated in the diverse natural surroundings of the Sierra de Montsec, at an elevation of 800 m. The ranch is home to Hilde and David, their 5 children, 19 horses and some goats. Guests can stay on a small campsite (6 pitches) or in a guestroom in one of the converted stables (4 quadruple bedrooms and 2 small dormitories for 6 guests each). All rooms have a view of the mountains. There is a bar/restaurant, a lounge with an open fireplace and a kitchenette for guests. Organic produce is for sale. There is room to stable your own horse - or you can reserve time on one of the ranch's horses. Rides last from three hours to several days. On the back of an Andalusian horse is the best way to see the magnificent mountains, fields and forests of oak, holm and pine.

The village of Benabarre is 1 km away. A maze of narrow streets leads you to Los Condes de Ribagorza castle, which offers a spectacular view of the Pyrenees. The farm is a good starting point for walks through beautiful cultivated fields and wild mountain scenery. The small Ciscar gorge is 6 km away. A bit further away, on the Catalonian border, is the Monrebes gorge.

How to get there

By car: From Lleida, take the N230 N to Benabarre. From Huesca, take the N240 to Barbastro, then the N123 to Benabarre. In Benabarre, follow signs for Viella. You will find 2 signs directing you to the Rancho. **Public transport:** Take a bus from Lleida or Huesca to Benabarre (3x daily). From there, it is another 1.5 km on foot. Or call to be picked up.

28. Allucant - Albergue Rural Ornitológico

> Javier Mañas Ballestín
> C/ San Vicente 1, 50373 Gallocanta, Zaragoza
> Tel/fax: 976-80 31 37 javier@gallocanta.com
> www.gallocanta.com
> Open all year Language: Ct, E, GB
> € Lodging pn: 2p 24-33; extra bed 7.50;
> Hostel pp: 7; B&B 8.50; half board 14.50;
> full board 19-22; breakfast 2.50; lunch 8;
> dinner 6; lunch-pack 6; VAT 7%

Hostel and surroundings

This newly built hostel (1995) is on the outskirts of the village of Gallocanta (elev. 1,000 m), near La Laguna de Gallocanta. This salt water lake has been designated as an internationally protected wetland area because of its incredible diversity of birds (more than 200 species). Almost all of western Europe's cranes (more than 70,000 birds) forage here on their yearly migration between their northern breeding

grounds and their winter territory in southern Spain. The best times to see large numbers of cranes are late October and late February. Javier is an avid bird watcher who organises field trips and teaches ornithology. Ask Javier to show you his large collection of books about birds.

The hostel provides simple but clean accommodation. There are 3 double bedrooms with adjacent bathrooms, 4 double rooms that share a common bathroom, and 2 large dormitories. There is also a meeting room (with audio-visual equipment), a bar and a restaurant (which offers vegetarian as well as meat dishes). There are bicycles for hire and a swimming pool in the village.

The lake, which dries up at the hottest time of year, is on a GR walking trail. Gallocanta has an ornithological museum. There are several villages near the lake that are worth visiting: Berrueco, Tornos and Bello.

How to get there

By car: From Zaragoza, take the N330 to Daroca (84 km). From there, take the A211 for 18.7 km (past Santed), where the road to Gallocanta (Z4241) turns off to the left. From there, it is another 3.6 km to Gallocanta. **Public transport:** information available on request.

29. Las Cardelinas

Carlos Pascual Lorrio
C/ Aurora 7, 50694 Pinsoro,
Zaragoza
Tel: 976-67 38 85 / 976-67 36 65
lascardelinas@terra.es
Open all year Language: E, GB
Lodging: €12.02; B&B €15.03; dinner €9.02

Farm and surroundings

The farmers at Las Cardelinas grow bio-dynamic vegetables and fruit - and guests are welcome to help out. There are chickens and ducks on the farm. The house is a traditional building surrounded by fruit trees. Your hosts serve meals daily during high season and whenever the number of guests permits. Otherwise, guests may use the kitchen for a small fee. The home-grown organic produce is for sale. There are 6 double bedrooms. Guests share the bathroom, living room and kitchen with the host family. Carlos and Ainhoa extend a warm welcome and organise Tai Chi classes, games, and activities related to organic farming and nature conservation. Bicycles are for hire. Pets are welcome.

Las Cardelinas is in the old part of Pinsoro; the woods and fields are only 20 m away. There are mediaeval villages in the area, as well as Roman ruins, Cistercian cloisters, lakes, gigantic rock formations (Bardenas Recles), 11th and 12th century Romanesque churches, castles and many other sights.

How to get there

By car: Pinsoro is on the Aragón-Navarra border. Take the road from Zaragoza to Logroño and Pamplona. You will reach Alagón after about 20 km. From there, head for Remolinos and Tauste (A126), Ejea de los Caballeros (A127), and Valareña (A125). Turn right towards Pinsoro. **Public transport:** Take a bus from Zaragoza to Ejea de los Caballeros, then another bus to Pinsoro.

Farm n° 19. Venta de San Juan

Navarra and Euskadi

Navarra

Navarra, in the western Pyrenees, is famous for 'Los Sanfermines', the yearly running of the bulls in Pamplona. Northern Navarra's Basque roots are visible everywhere: in the landscape, in the people, and particularly in the villages of the Baztán valley. The Ebro valley, which forms the border between Navarra and La Rioja, shows clear traces of Aragón's and La Rioja's influence on the local architecture and culture.

Navarra has countless walking trails that lead you through both unspoilt nature and quaint villages. Navarra has an abundance of deep oak and beech forests, mediaeval towns, incredible panoramas and steep mountains with pine trees that have taken root in the most unlikely places. Navarra's largest agricultural area is La Ribera, where the chief crop is wine grapes - particularly the Tempranillo and the Garnacha.

Roncesvalles, near the French border, is a historic stop on the pilgrimage to Santiago de Compostela. In the east, where Navarra borders on Aragón, you will find the Valle de Roncal, a lush valley grazed by the sheep that have made the village of Roncal famous for its cheeses. Navarra's north-east is dominated by the Pyrenees. Some of Spain's biggest forests are here, in particular the Bosque de Irati, where the nearly extinct brown bear is sometimes spotted.

Basque Country

The Basques, farmers and sailors by tradition, are a proud people who have been struggling for greater independence for decades now. They have their own language: Euskera. Euskadi, or Basque Country, stretches from the Pyrenees in the north-east to the dry Spanish Meseta in the south and Cantabria in the west. The greatly varied landscape is dissected by numerous rivers and by the Cordillera Cantábrica, a rough but verdant mountain range with peaks reaching 1,550 m.

Farm no 43. Olabarrieta Beheko

The northern coast of Basque country is some 200 km long. Here, you will find picturesque fishing villages, river deltas, sandy beaches and rocky cliffs. The coastal province of Biskaia is home to the Urdaibai wetlands. Gipuzkoa has the most mountains and deep, green valleys in all of Basque country. It is home to two large nature reserves: Sierra de Aralar (11,000 ha) and Sierra de Aitzgorri (18,000 ha). Amidst the wooded mountains of the interior, you will find a wealth of castles, monasteries and old family farms where traditional mixed agriculture is practised.

Typical scenes in this beautiful countryside include cows and sheep grazing the mountain slopes, steep meadows mown with a scythe, and haystacks in the farmyards. Further south are the dry plains of the famous Rioja wine region (Araba province). The region abounds with *bodegas* and historical villages such as Biasterri (Laguardia).

All farms in Basque country are members of Nekazalturismoa (see page 103).

Geese at farm nº 35. Naera Haundi Baserria

30. Primorena

Rodrigo Barricart
Calle San Pedro 16, 31481 Meoz, Navarra
Tel/fax: 948-39 03 06
Open all year Language: E, F, GB
B&B pn: 2p €33.50; extra bed €9.01;
lunch & dinner €10

House and surroundings

Primorena is a village with only 7 permanent residents. The owner and his friends rebuilt his 300 year-old house stone by stone, re-using nearly all the original construction materials. Heat from the living/dining room fireplace is channelled throughout the building. Doors and furniture are either home-made or second-hand. This relaxing house has a sauna, outdoor swimming pool, a serene and sunny piano room and a family room where tasty vegetarian, organic meals are served. In summer, courses in yoga, music therapy and dance are available. There is room in the garden for 2 small tents. Your host, Roberto, is involved in a solar and wind energy project and will soon be purifying the pool's water by environmentally friendly means. No smoking or alcoholic beverages permitted in house or garden.

The village has a 13th century Gothic church, la Iglesia de San Pedro. From Primorena you can look out over the grain fields of Lónguida valley. Just 10 km away are the Sierras de Arratxaba and Zariquieta nature reserves, at the foot of the Pyrenees. There are also several gorges to the east, such as Foz de Lumbier and Foz de Arbayún, with spectacular natural beauty, many otter colonies and birds of prey such as the lammergeyer and several owl species. You can swim in the Pantano de Yesa reservoir, just over the border in the region of Aragón.

How to get there

By car: NE of Pamplona, take the NA150 towards Lumbier. Approx. 4 km after Aós, turn left towards Meoz. Drive into village. It is the first house on the corner of the street leading to the church. **Public transport:** Train to Pamplona, then take the bus that goes to Lumbier. Ask to be dropped off at the exit to Villanueva de Lónguida and Meoz (another 2.5 km).

31. Urruska Baserria

Joxepi Miura Iribarren
Apartado 65 - Barrio Beartzun
31700 Elizondo, Navarra
Tel/fax: 948-45 21 06 urrusca@teleline.es
www.baztan-vidasoa.es
Open Jan-Oct Language: Eu, Language: E, F
Lodging pn: 2p €31; breakfast €3; lunch
€13; dinner €10

Farm and surroundings

Urraska is a traditional cattle farm with Pyrenean cows and indigenous Latxa sheep. The animals graze the surrounding mountain meadows in the summer. In winter, they eat home-grown corn and unsprayed cattle feed from a local co-op. The farm has won an interior decoration award; everywhere you look, there are traditional farm equipment, antique furniture heirlooms and special utensils - such as a tool used for roasting apples on an open fire. Guests can tour the working farm. Every guest room is named after the mountain visible from its window; naturally it is a must for guests to climb 'their' mountain! The hosts serve traditional dishes made from ingredients produced on the farm. You are welcome to help out with the vegetarian cooking. Traditional cheeses are sold on the farm. This is a beautiful area where you can take long walks and visit villages such as Elizondo, where life revolved around cattle raising for centuries.

How to get there

By car: From Pamplona take the N121-A to Mugairi. There, turn right on to the N121-B to Elizondo. In Elizondo, drive past the church, then turn right on the NA2596 towards Beartzun (another 10 km). Follow signs to Urruska. After passing the gate, take the unpaved road bordered by stone walls (850 m). **Public transport:** Train or bus to Pamplona (Iruñea), from there take the bus to Elizondo, get off at the Baztanesa bus stop. Take a taxi (10 km).

32. Hotel Peruskenea

José María Astiz
31866, Beruete (Basaburua Mayor)
Navarra. Tel: 948-50 33 70
Fax: 948-50 32 84
peruskenea@peruskenea.com
www.peruskenea.com
Open all year Language: Eu, E
€ pn Lodging: 2p 66.11(ls)-72.13(hs);
breakfast 7.21; lunch & dinner 16.83; B!

Farm and surroundings

This family farm, situated in the foothills of the Sierra de Aralar, has been converted into a comfortable country inn. The cattle once bred here have given way to organic fruit orchards. Most of the beehives are kept in the woods nearby. Owner José-María is currently working with the 12 villages of Basaburua municipality to open up a shop selling local organic produce. He aims to sell his fruit and honey there,

as well as locally made cider and products from a nearby dairy farm now in transition to organic farming. Surrounding the inn is a garden which borders on 7 ha. of oak and beech woods. If you take an early morning stroll into the surrounding hills you can catch a spectacular view of the dew-covered landscape. The inn has two comfortable sitting rooms with fireplace. Nearby there are marked footpaths which lead you past unspoilt natural surroundings, ancient burial mounds, a deserted railway track (the Plazaola line), ruins of old bridges and tunnels carved out of the mountains. Starting in the Spring, many traditional festivals are held in this area: carnival in Alsatsu, a medieval market in Leitza, a herders' festival in Uharte-Arakil, and local folk games such as wood-chopping and stone-lifting competitions in several villages.

How to get there

By car: On the A15 from Pamplona to San Sebastián, get off at the Urritza-Latasa exit. Take the NA4110 to Jauntsarats, where you turn left on to the NA4112. At the 'Peruskenea' sign, turn on to the unpaved road and follow for 1 km. **Public transport:** Train from Pamplona or San Sebastián to Irurzun. Then take a bus to Latasa. Call ahead to be picked up.

33. Casa Etxeberri

Josefina Iturralde & Jesús Mari Ollo
31869 Goldaratz, Navarra
Tel: 948-50 32 38
Open all year Language: Eu, E
Self-catering: €108.20 weekend; €312.52 pw;
lunch & dinner €9.01; VAT 7%

Farm and surroundings

Casa Etxeberri is a farmhouse situated in the Imotz valley. Nearly all of the village's 35 inhabitants are involved in cattle raising. Josefa and Jesus-Mari were the first to switch over to organic farming. They raise Basque Latxa sheep. The cheese, meat and cider they serve with dinner are all produced on the farm. Josefa and Jesus-Mari are proud to show you their farm, demonstrate how they make cheese or guide you to a nearby megalithic burial site and caves. Guests are welcome to do chores on the farm. There are 3 spacious, simply furnished apartments and a large sitting/dining room where meals are served. The garden has a cosy sitting area with a view of the surrounding fields and mountains.

Every year on September 8, Goldaratz celebrates its patron saint. The village's Romanesque church is worth a visit. You can also take sightseeing day trips in the surrounding area. Three long-distance hiking trails pass within 10 km of the farm. Pamplona (26 km away) has several Gothic and Romanesque churches and great cafés and restaurants on the narrow streets around the Plaza del Castillo. The city holds San Fermín celebrations from August 6-14.

How to get there

By car: From Pamplona take the A8 towards San Sebastián. Exit at Latasa-Urritza. From there it is another 2.5 km to Goldaratz. **Public transport:** Take a train or bus to Pamplona, then a train to Irurtzun, where you can take a taxi. Or take a bus to Latasa and walk the last 2 km.

34. Aldarreta

Maritere Lardizabal & Pedro Urdangarin
20211 Ataun, Gipuzkoa
Tel: 943-18 03 66
Open all year Language: Eu, GB
Lodging pn: 2p €27.05; breakfast & lunch-pack €2.40; lunch & dinner €8.41

Farm and surroundings

This 500 year old farm is splendidly situated at the edge of a nature park and the Sierra de Aralar mountains. It is a high, isolated spot amid lush meadows and woods, with a panoramic view of the mountains. The silence is broken only by the cowbells on the 24 cattle that graze the steep grasslands in the summer. The beef produced here is a protected regional product, with the Kalitatea certificate of origin. The farm has a large traditional vegetable garden where no pesticides are used. Maritere uses her fresh produce to cook meals for guests (on request). The four rooms, which share two luxurious bathrooms, are outstanding. There is a large sitting/dining room with open fireplace, a covered terrace and an open terrace with a wonderful view. Guests are welcome to use the fully-equipped, modern kitchen. From the farm, you can walk directly into the woods and the mountains. The owners, who speak Basque and Spanish, have acquired an English dictionary to help them communicate with guests. The village (2.3 km) has a swimming pool and a farmers' market where you can do your shopping. Walking tours with guides are organised nearby. There are also horses for hire.

How to get there

By car: From San Sebastián (Donostia), take the N1 to Vitoria (Gasteiz). Take exit 419 to Beasain. Then take the GI120 to Lazkao, continue on this road for 3 km past Lazkao, to Ataun. Turn right at the 'Nekazalturismoa' sign. Take this winding mountain road for another 2.3 km, following the wooden signs to Aldarreta. **Public transport:** Train or Eurolines bus to San Sebastián (Donostia). From there, take the local train and get out at Beasain. Take a taxi, unless you have arranged in advance to be picked up.

35. Naera Haundi Baserria

Jesús María González &
Angela Linskey
20269 Abaltzisketa, Gipuzkoa
Tel: 943-65 40 33
naerahaundi@euskalnet.net
Open all year Language: E, GB, F, Eu
Self-catering pn: €48.80; breakfast €3;
VAT 7%

Farm and surroundings

This old farm and cider factory is in a scenic, wooded valley. The house is at the end of a road, and the only noises to be heard are those of the farm, the animals and birds. The cider factory was built in 1706 and is now one of the historical monuments of the Basque province of Gipuzkoa. Owners Jesús María and Angela make a living by selling home-made organic jam from the many fruit trees (apple, pear and

kiwi, for example) and berry bushes surrounding the house. The vegetable garden also produces ingredients for their jam. You can purchase fresh cheese (must be ordered in advance), jam, milk, fruit, eggs, pies and chocolates. Guests are welcome to help harvest the fruit: the kiwis, for instance, are picked in November. The farm also has chickens, geese and a few goats. The guest lodgings are sparingly furnished. There are two apartments with bath and kitchen for up to four guests each. In the winter, the rooms and water are heated by wood chopped on the premises. Guests have access to a terrace and garden. Safe play area for children.

The owners organise local walking tours with Jesús María as your guide (EE24 including lunch pack). There are also mountain bike paths (5 km away) and riding paths (20 km). Cattle herders tend their Basque cows in Aralar nature park (3 km), home of Mount Txindoki (elev. 1346 m). This park features centuries-old menhirs. Higher up in the mountains, you may come across vultures.

How to get there

By car: From San Sebastián (Donostia), take the N1 towards Vitoria (Gasteiz). Approx. 2 km past Tolosa, turn off towards Alegia. Drive into Alegia, cross bridge and take the first right turn, uphill to Abaltzisketa (GI3670). After 7 km, at a sign reading 'Nekazalturismoa', turn right towards Sasiain. After 100 m (at the first farm), turn right again and drive 900 m downhill to the road's end. **Public transport:** Train or Eurolines bus to San Sebastián, from there take the train to Tolosa (13 km) or Alegia (5 km). From either locality, take a Tolosaldea bus 1 km to the Sasiain bus stop.

36. Sarasola-Zahar

Ana María Múgica
Barrio Zehatz, 20749
Aizarnazabal, Gipuzkoa
Tel: 943-14 77 74
sarasolagro@yahoo.es
Open all year Language: E, F, GB
Lodging: 2p €24.04

Spain

Farm and surroundings

This is a pretty farm in a remote, hilly landscape. It is situated at the end of a small road about 10 km from the coast, near San Sebastián. Ana is a hospitable proprietor. In 1976, her parents bought the farm, which had been a local school until the 1950s. It also served as a bakery, supplying the entire surrounding area. Ana's parents began raising cattle there; in 1990 they made the transition to organic fruit growing. It is now run by Ana and her father. The farm houses chickens, geese and horses. The apples in the orchard (1.5 ha) are crushed to make cider. Part of the land is grazed by cows from the neighbouring farm. Ana has three children. During the summer months, you can purchase vegetables from the garden. The spacious kitchen is available for use. The lodgings are sparingly furnished. There is one shared bathroom for every two lodgings. Surrounding the house is a large garden with many fascinating nooks and crannies for children to explore. Sitting on the terrace, you can listen to the sounds of nature while enjoying views of the sea and the farmland. Offers to help on the farm are usually welcome.

You can take walks, ride and go to the beach. Bilbao (Guggenheim Museum) and San Sebastián are a 45 minute drive by car.

How to get there

By car: Take the A8 or N634 San Sebastián-Bilbao, take the Zarautz exit. In Zarautz, turn left on the GI2633 towards Aizarnazabal/Zestoa. In Zubialde, just before the Bar 7 cafe, turn left and drive uphill following the 'Nekazalturismoa' sign. Then follow the wooden signs to Sarasola-Zahar (3 km) until you reach the farm.
Public transport: Train or Eurolines bus to San Sebastián. From there, take the train (Eusko Tren) to Zarautz. Then take the Guipuzcoana bus to Aizarnazabal. Get off at the bus stop 'Bar 7 de barrio Zubialde'. Walk the remaining 3 km. Alternative: take a direct bus (Guipuzcoana) from San Sebastián to Iraeta. Call farm for a pick-up.

37. Baserri Arruan Haundi

Kontxi Argaia & Willem Eisenaar
Barrio Lastur, 20820 Deba
Gipuzkoa. Tel: 943-60 37 04
lastur20@euskalnet.net
Open all year except Dec 20th-29th
Language: E, NL, GB, F, D, Eu
€ *pn Camping: tent 3.15; caravan 3.50; adult 3.27; child 2.45; Self-catering: 63.11(ls)-84.14(hs); Groups: 78.13; breakfast 3; VAT 7%*

Farm and surroundings

The farm, a former iron foundry, stands in a peaceful mountain valley near the Basque coast. There are cows, horses, chickens and a large herd of sheep that provide milk for home-made organic cheese. Owners Willem and Contxi demonstrate their cheese-making methods and also show guests how to comb and spin wool and weave fabric. Their daughters play Basque music on a *soinu* (a basque accordion) and *pandoreta* (tambourine). Guests may help tend the farm animals. An organic grocery store on the premises sells fruit, vegetables, home-made cheese, jam, cider and buttermilk.

Between the stables and the farm is the guest cottage, with kitchen, bathroom, sitting room and three bedrooms for up to seven people. Next door is 'The Hostel', with two rooms and bunk beds for up to 14 guests. Both houses

42

can only be rented out as a whole. Hostel guests and campers use the farm's toilets, showers, sitting room (with fireplace) and kitchen. All guests should bring their own towels; hostel guests should also bring their own sleeping bags. The surrounding area offers many walking, cycling and caving opportunities. Plenty of tourist information is available. Guests can choose from several beautiful beaches (approx. 12 km away).

How to get there

By car: San Sebastián - Bilbao on the A8 or the N634. Get off the A8 at exit 13 (Itziar/Deba). Turn left, briefly heading towards Itziar on the N634, then right towards Itziar/Lastur, and immediately left again towards Lastur. Follow road past an industrial park until the next three-forked road. Turn right, on to the GI3210 towards Elgoibar. Follow the signs for Lastur, and after a few km turn right towards Lastur on the GI3292. Drive through the hamlet of Lastur, into the valley, and after 2 km - at the second 'Nekazalturismoa' sign - turn left on to the farm. **Public transport:** Train or Eurolines bus to San Sebastián, from there the bus to Itziar (Iciar). Call ahead to be picked up there. Alternative: Train from San Sebastián to Deba, taxi (approx. €12) to the farm.

38. Mendiaxpe

María Jesús Mendia
C/ Salsamendi 22, 01250 Araia, Araba
Tel/fax: 945-30 42 12 Mob: 646-10 45 84
Open all year Language: E, Eu, GB
€ pn Lodging: 1p 24; 2p 27; B&B 1p 27;
2p 42; extra bed 9; breakfast 3

House and surroundings

Mendiaxpe is a former farmhouse in Araia (elev. 600 m) wedged between the Sierra de Aizgorri, Urkilla and the Sierra de Urbasa. The owners restored the house, which was built in 1896, to its former beauty and decorated it with restored family furniture. There are 3 double bedrooms with *en suite* bathrooms and a big eat-in kitchen for guests to share. You may use produce from the vegetable garden in preparing your meals. Chickens and ducks live under the trees in the large orchard. Honey is harvested in October, from the 15 beehives on

the property.

The name Mendiaxpe means 'at the foot of the Stone Mountain', a reference to Mount Araz which rises 1,445 m from the back of the house. A path that starts there leads you to the peak in about 2 hrs. On the slopes, you may come across shepherds and their sheep. Bicycles are for hire nearby and information about nature walks, mountaineering and other outdoor activities is readily available. The towns of Araia and Agurain (Salvatierra) offer arts and culture.

How to get there

By car: Araia is approx. 18 km W of Alsazua. On the N1 W (Iruñea-Vitoria (Gasteiz)), take exit 385 to Araia. Once there, follow the signs to Nekazalturismoa Mendiaxpe. **Public transport:** Take a train to Vitoria (Gasteiz). From there, take a bus (5x daily) or a train to Araia. Walk the last 2 km, or take a taxi, or call to be picked up.

39. Erletxe

María Arrate Aguirre
Rua Mayor de Peralta 24-26, 01300 Biasteri (Laguardia), Araba
Tel/fax: 945-62 10 15 Mob: 657-79 93 11
erle.etxe@euskalnet.net
Open Mar 15th-Dec 15th Language: E, F, GB
Lodging pn: 2p €48; breakfast €3; VAT 7%

Pension and surroundings

The perfectly preserved mediaeval town of Laguardia is now a protected Spanish monument. The SW corner of the 13th century city wall is where you will find the beautifully restored and decorated home of Maria and her husband. The house looks out over expansive Rioja vineyards, with *bodegas* that are open to the public, except during summer break. Hostess Maria can tell you a great deal about local history. She is a beekeeper who prepares her honey at home. This is why her pension is named 'Erletxe' - Basque for beehive. She also owns 300 walnut and fruit trees and makes her own preserves, which she serves with whole grain bread for breakfast. Guests may only use her kitchen to prepare baby food.

There are 5 double bedrooms, 3 with their own shower/ bathroom. The owners occupy the top floor.

The flat countryside is ideal for cycling. Sierra de Toloño (only 10 km away) has several peaks of over 1,300 m and good footpaths. An archaeological dig nearby has revealed an approx. 3,500-year-old settlement (La Hoya) and several megalithic burial sites. Two km away are the Carralogroño and Carravalseca lagoons and Prao de la Paul, a sanctuary for aquatic birds.

How to get there

By car: Take the N111 from Pamplona (Iruña) SW to Logroño. There turn on to the A124 to Laguardia (Biasteri). As you enter town, turn left. After a few hundred metres, there is an intersection where 5 roads meet - take the 2nd road to the right. **Public transport:** Train or bus to Logroño or Vitoria. From there, take a bus to Laguardia (Biasteri).

40. Uxarte

Gotzone Larrañaga & Javier Padilla
Barrio Untzilla, 01169 Aramaio (Ibarra)
Araba. Tel/fax: 945-44 51 46
Open all year Language: E, Eu, D, F, GB Lodging pn: 2p €36.50; breakfast €3.50; dinner €10

Farm and surroundings

For five generations now this stately farm, which stands in a lush landscape at the edge of a small village in Aramaio (a region known as Little Switzerland), has belonged to Gotzone's family. Gotzone is a friendly Basque farmer who lives there with her husband and four children, her mother and sister. On the farm, there are five cows (milked by hand), sheep, chickens, geese, turkeys and other animals. The farm has a large organic vegetable garden and many different fruit trees. The five impeccable double rooms with bath/shower are in the former stables. Meals contain as many locally-grown ingredients as possible and vegetarian dishes are available. Gotzone can supply you with ample information regarding sightseeing and interesting day trips in the vicinity.

Untzilla is near Urkiola nature park, and 20 km from Aizkorri nature park with its wonderful footpaths. There is a beautiful reservoir where you can go swimming and canoeing just 10 km from the hostel.

How to get there

By car: Travelling from San Sebastián to Bilbao on the A8, take exit 15 on to the GI627, towards Vitoria. In Aretxabaleta, go towards Ibarra, follow this road until Untzilla, from there follow 'Nekazalturismoa' signs. **Public transport:** Train or Eurolines bus to Vitoria, from there the bus to Arrasate, get out in Aretxabaleta. Walk remaining 3 km or call to be picked up.

41. Guikuri

Pilar López Conde
Guikuri 1, 01138 Murua, Araba
Tel/fax: 945-46 40 84
guikuri@bezeroak.euskaltel.es
Open all year Language: E, GB
€ pn Lodging: 1p 30; 2p 39; breakfast 4;
dinner 15; VAT 7%

Farm and surroundings

This superbly restored farm was built from wood and natural stone in 1850. It is located in the Gorbea nature park on a narrow road bordered by beautiful stone walls. The farm belongs to 2 young families with small children. They do their best to maintain an environmentally friendly lifestyle. They own horses, chickens (for fresh eggs) and four pigs. The farm has an organic vegetable garden and a shop which sells self-restored antiques, and arts and crafts items.

The 5 guest lodgings, some of which are in the former barn, share a common kitchen/dining area and a spacious, comfortable sitting room with a fireplace. In front of the lodgings there is a lovely terrace and a large garden. On request, guests are served home-baked organic whole grain bread and cake for breakfast and a delicious dinner. Bicycles and canoes are for hire on the farm. Directly from the lodgings, you can take a 1-hour walk through a splendid beech forest to the Mairuelegorreta cave, an underground paradise of limestone rock for-

mations where according to local legend a treasure lies buried. You can swim and canoe in the reservoir in nearby Legutiano.

How to get there

By car: From Vitoria (Gasteiz), take the N240 N until approx. 3 km past Legutiano. Turn left on to the A3608 towards Etxagüen and Gopegui. After about 6 km, at the village of Murua, turn right and follow the Nekazalturismoa and Guikuri signs. **Public transport:** Train or Eurolines bus to Vitoria, from there a taxi or call in advance to be picked up. On Thursdays and Saturdays, there is a bus from Vitoria that passes through Murua.

42. Ibaizar Baserria

Koldo Mendioroz
Sojo 4, 01478 Sojo-Zollo, Araba
Tel/fax: 945-39 66 86
Mob: 652-77 65 55
info@ibaizar.com www.ibaizar.com
Open Jan 6th-Dec 31st Language: E, Eu
€ pn Lodging & self-catering: 2p 54.09;
breakfast 3.91; lunch 22.84;
dinner 9.01-21.03; lunch-pack 3; VAT 7%

Farm and surroundings

Ibaizar, a beautifully restored 18th century manor house, sits on a hilltop (elev. 400 m) at the foot of the Sierra Salvada, overlooking the Aiala valley. Several years ago Josu, his wife and another couple started restoring the house. They transformed what was no more than a ruin into a gorgeous manor house with 3 double bedrooms and 2 self-catering apartments, each with its own ambience and special features. You probably won't want to use the apartment's kitchen once you have tasted what the hosts serve at their own table! They use home-grown organic vegetables and fruit for these delicious meals. One apartment is wheelchair accessible. There is a common room with an open fireplace. Fresh home-grown and local produce is for sale when in season.

The Aiala valley is truly an ecological paradise, shielded from mass tourism. The mediaeval town of Artziniega, 4 km away, has a walled inner city with many churches worth visiting. Right across the border of Burgos province is the Monumento Natural Monte Santiago. Jagged limestone cliffs form a spectacular natural amphitheatre, which is the backdrop to the 300 metre-high Salto del Nervión waterfall. The forests and caves are home to eagles, wolves, vultures and deer. If enough people sign up, the owners organise 1-day excursions to this nature reserve. The city of Bilbao (Guggenheim Museum) is 30 km away.

How to get there

By car: From Bilbao, take the BI 636 to Sodupe. From there, take the BI 504, then the BI 2604 to Artziniega. Once there, take the A3624 via Sojoguti to Sojo-Zollo. **Public transport:** Take a bus from Bilbao to Artziniega. Call to be picked up there.

43. Olabarrieta Beheko

Asier Ulibarri
01409 Okondo, Araba
Tel: 945-89 81 46
olabarrietabehe@euskalnet.net
www.euskalnet.net/olabarrietabehe
Open all year Language: E, Eu
€ pn Lodging: 2p 42.07; B&B 2p 50.49; extra bed 14.73; breakfast 4.21; lunch 9.02-24.04; dinner 9.02-18.03; VAT 7%

Farm and surroundings

Olabarrieta Beheko is a traditional Basque family farm which is currently in transition to organic farming: the 300 sheep will be certified organic by June 2002. The vineyards, which will be certified at a later date, produce grapes for *Txakoli*, a traditional Basque white wine. The village of Okondo (elev. 150 m, population 800) is situated in a valley surrounded by mountains 400 - 1,000 m high. The farm has 6 double rooms with *en suite* bathrooms, 2 lounges and patios on all sides of the house with a view of the fields. Breakfast is served with fresh eggs and fresh bread from a wood-burning oven. Dinner consists of traditional Basque fare prepared with home-grown and home-made products. Guests can study sculpture and drawing, watch how the traditional bread is made, and help the owner and his 3 dogs herd the sheep. The family - Angel Mari, Josefi and their children, Asier and Iker - have produced an excellent video of traditional farming practices, such as ploughing with oxen and making charcoal. Farm products and wood carvings are for sale. The nearest place to swim is in the village, 1.5 km away.

The area is ideal for walks and bike rides through green fields and mediaeval villages with local museums. The Parque Natural de Gorbea, a Karst (limestone) range that is home to many special plant and animal species, is some 20 km away.

How to get there

By car: From Bilbao, on the A68 S (to Madrid/Barcelona), take exit 3 to Llodio. From there, head for Okondo. **Public transport:** There is a Bilbao-Okondo bus every 2 hrs (bus stop is 1 km away). Or take a train from Bilbao to Llodio, then a taxi for the remaining 7 km.

44. Baserri Amalau

Gabriel Vázquez & Puri Adrian
Barrio Ipiñaburu, 48144 Zeanuri
Biskaia. Tel/fax: 946-31 71 79
artasolo@hotmail.com
Open all year Language: E, F
€ pn Self-catering: 45-102; 300-630 pw;
extra bed 9; breakfast 3; VAT 7%

Farm and surroundings

This organic farm is set in a hamlet of four farmsteads. There are two nature parks nearby - Gorbea (100 m from the farm) and Urkiola (15 km away). Owner Gabriel is a botanist and kiwi grower. The farm's natural surroundings inspired him to write two books: *Medicinal Plants of the Basque Country* and *The Magical Power of Plants*. In a room filled with the scent of fresh and dried herbs, Gabriel gives courses in medicinal botany, cosmetics, permaculture and ecology. The 2 apartments downstairs have a rustic decor and old Basque wood furnaces. One apartment is decorated in forest patterns. On request, Gabriel takes groups on botanical excursions. Breakfast can be ordered. Other meals (fish or vegetarian) are available for groups of approx. 10 guests.

Help on the farm is especially welcome during the November kiwi harvest. Bicycles are for hire and there is plenty of information on nearby cycling and walking tours. In the summer, nearby villages have open air markets that sell local organic produce and old Basque arts and crafts products. Near the farm you will find old mills, herds of indigenous Latxa sheep, excellent caving sites, rare wildlife, plants and summer village festivals. The farm is within 40 km of Bilbao (Guggenheim), Vitoria and the ocean.

How to get there

By car: San Sebastián (Donostia) towards Bilbao on the A8/E70 or N634. Take the exit to Amorebieta. In Amorebieta, take the BI635 towards Lemoa and go all the way to the N240. Head south towards Vitoria (Gasteiz). After 12 km, in Zeanuri, turn right and drive over a small bridge towards Barrio Ipiñaburu. Take this narrow road 6.5 km into the mountains until you reach the farm (follow Nekazalturismoa signs). **Public transport:** Train or bus: via Irn to Bilbao. From there, take the Bizkaibus A-3917 Bilbao-Lemoa-Zeanuri, which runs from 6:15 a.m. - 10:15 p.m., twice hourly on weekdays; once per hour weekends and holidays. Get off at Zeanuri. Call in advance to be picked up.

45. Urresti

María & José Mari Goitia
Barrio Zendokiz 12
48314 Gautegiz de Arteaga
Biskaia. Tel/fax: 946-25 18 43
urresti@nexo.es
Open all year Language: Eu, E, GB, F
€ pn Lodging: 2p 38(ls)-44(hs); extra bed 12;
Self-catering per unit: 58; breakfast 4; VAT 7%

Farm and surroundings

In 1988, María and Jose Mari's dream came true. They bought a dilapidated seventeenth-century farm and planted their organic pear orchard. The house nestles in the hills of UNESCO's Urdaibai biosphere reserve, just 4 km from the coast and the fishing village of Erlantxobe. There are many farm animals and a large vegetable garden. The rooms are in the former stables of the old farm; in the hallways, the walls are lined with framed photographs of guests who enjoyed their stay. The rooms are decorated *al estilo María*; old and new furnishings are combined. The kitchen may be used for an added fee. Two nice apartments are also for hire.

Your hosts can arrange walks in the surrounding area and guided tours of the Guggenheim museum. Sometimes, Jose Mari brings guests along on fishing trips to catch anchovies in the Cantabrian Sea. These are said to be the best anchovies in all of Europe, due to the water temperature and the rough seas. The Urdaibai biosphere reserve (220 square km) is great for nature lovers. Walks down the coast take you past picturesque fishing villages, lovely beaches and steep cliffs.

You also come across rivers, woods and rolling farmland. Paddle down the Ría de Gernika by canoe and observe the wetland wildlife where many migratory birds stop in late summer. Or visit painter Ibarrola's Enchanted Forest, the caves of Santimamiñe and Gernika.

How to get there

By car: Take the A8/E70 from San Sebastían towards Bilbao, get off at exit 18 (Gernika). In Gernika, take the BI638 towards Lekeitio. Drive through Gautegiz de Arteaga. At the next intersection (after a 3.5 km incline uphill) get on the BI-3237 towards Elantxobe. Continue for exactly 1.2 km after the intersection, passing another 'Nekazalturismoa'-house. Turn right (downhill) immediately after km marker 41.
Public transport: Train or Eurolines bus to San Sebastián or Bilbao. Then train to Gernika; here Bikzaibus to Lekeitio via Elantxobe. Get out at Basetxetas bus stop, 0.3 km from farm.

Farm n° 31. Urruska Baserria.

Cantabria, Asturias and Galicia

Cantabria

Cantabria is a patchwork of several different landscapes in a relatively small area. The coastal region in the north, the high mountains of the Cordillera Cantábrica, and the valleys of the seven rivers cutting through the region - each have their specific characteristics. Many of these special landscapes have been declared protected areas.

Cantabrians have always been headstrong and independent. Even the cave dwellers of Altamira were known for their fighting spirit! The Romans, and later the Moors, failed to conquer the Cantabrian rebels. The ports have always been of great importance for overseas trade and emigration. Nowadays, the economy is largely based on the service industries, and to a slightly lesser extent on agriculture, cattle breeding and fisheries.

Asturias

Asturias is the mountainous area between Cantabria in the east and Galicia in the west. For centuries this was the seat of the Christian empire, leading the struggle against the Muslim empire of southern Spain. Asturians claim the famous Picos de Europa as a symbol of their identity, even though this mountain range also stretches into other regions.

Picos de Europa

The impressive Picos de Europa - recently declared a national park - start just 25 km from Spain's northern coast. The mountains are partly in Asturia, in Cantabria and in Castilla-León. The park measures 40 km in width and is bordered by two deep river valleys (stretching from the river Sella to the river Deva). Its highest summit is Naranjo de Bulnes (2519 m). There are 28 villages in the mountains. Some, like Bulnes, can only be reached by a cliff path high above the river.

The natural beauty of the area can be compared to the Swiss Alps. Lovely green pastures with grazing sheep alternate with jagged rock formations, crystal clear lakes and snow-capped peaks. At high elevations it can

*Beach near Ber,
A Coruña, Galicia*

snow any time of year and the weather can be treacherous. It rains frequently in these mountains. One of the park's main attractions is its birds and other wildlife. Griffon vultures and kestrels hover over the mountains that sheep and mountain goats call their home. There are wolves, and even bears here: it is estimated that there are 60 of the rare indigenous bear, the *Ursus Ibericus*.

There are countless footpaths and long distance trails that lead you through river valleys, woods and the rough terrain of the high mountains. The routes vary in difficulty. The 11 km walk through the Garganta de Cares, also known as the 'Garganta Divina' (Divine Gorge), is spectacular. You walk from Caín to Puente Poncebos, following the twisting course of the river Cares.

The main park entrances are in Potes, Cangas de Onis (both have tourist information centres), in Arenas de Cabrales and Posada de Valdeón.

Galicia

Galicia, in Spain's far north-western corner, was a Celtic colony for five centuries. In the north and west, the landscape is dominated by the many river inlets that cut deeply into the land between the mountains. Further inland, the rolling green hills form an almost pastoral scene. In the south, Galicia borders on Portugal. The *lingua franca* of the region is 'Gallego', a language related to Portuguese. Galicia's capital, Santiago de Compostela, has been a world famous place of pilgrimage since the Middle Ages.

46. Albergue Rural La Tejedora

Yolanda Gutierrez & Robert Llurba
Barrio Casavieja 6, 39869 Ojebar (Rasines),
Cantabria. Mob: 646-14 91 05
alberguetejedora@hotmail.com
www.geocities.com/albergue_latejedora
Open all year Language: E, Ct, Ga, GB, F
€ pn Lodging: 2p 48.08 Hostel pp: 13.22;
breakfast 1.80; lunch 9.01; dinner 6.01

Hostel and surroundings

La Tejedora, or The Weaver, is in the small farming community of Ojebar ('Eye of the Sea'). The hostel stands at an elevation of 310 m, overlooking a rolling landscape of pastures and forests, dotted with villages and the occasional shepherd's hut. In the distance you can see the Ría de Treto inlet and the Sierra del Hornijo. Your hosts Robert and Yolanda used natural building materials to convert an old village house into their cosy hostel. They promote culture and nature conservation, and in the village they repair old fountains and washing basins. They also help to reforest the area, using indigenous trees.

The hostel has 2 double rooms with their own bathrooms and toilets, and 2 dormitories (one with 8 bunk beds, and another with 9). The common room, which has a balcony and large windows, is used for teaching various classes including Tai Chi and relaxation exercises. Guests may use the patio and garden. Meals are prepared with home-grown and local organic produce.

La Tejedora is an ideal starting point for outdoor adventures, especially for those interested in caves. Several menhirs and dolmen as well as Monte Ranero are within walking distance. A bit further away is the Reserva Natural de las Marismas de Santoña y Noja, a complex ecosystem of salt water flora and fauna and an important foraging ground for migratory birds. The Parque Natural Collados del Asón has spectacular limestone rock formations and a splendid waterfall at the source of the river Asón. The village of Rasines (4 km away) is known for its square bullfighting arena.

How to get there

By car: On the E70 Bilbao-Santander, take exit 173 to Burgos. Drive for about 12 km to Rasines. Once in the village, follow the sign to 'Ojebar 4 km'.

By bike: From Bilbao to Balmaseda (Karrantza) on the BI636. Three km before Balmaseda, turn on to the BI630 to Laredo. After about 12 km, turn off and continue on the BI630 to Rasines and Ojebar. Public transport: Ojebar is 16 km SE of Laredo. From Bilbao take the Termibus to Laredo, then a bus to Rasines. Or take a local FEVE train from Bilbao to Gibaja (6 km away). From there, take a taxi or call to be picked up.

---

47. Agroturismo Muriances

Hermanos Pérez Gutiérrez
Vilde, 33590 Ribadedeva, Asturias
Tel: 985-41 22 87
Open all year Language: E, GB
Camping, lodging, self-catering & meals:
prices on request

Farm and surroundings

This partly restored building, a former flour mill, is surrounded by meadows. The house is built into the side of a large rock. Across the road is the Deva river, which flows into the sea

3 km downstream. The owners make a living from cattle raising and growing pesticide-free fruit for jam production.

Part of the building has been converted into 5 spacious guest rooms. Tents may also be set up in the front yard between the fruit bushes. Campers may use the bathrooms and showers in the house, as well as the 2 cosy common sitting rooms. A glass wall in the downstairs sitting room offers a direct view of the cows in the shed. Your host offers breakfast and dinner, but you may also use the large kitchen to prepare your own.

Attractions nearby include the Picos de Europa (10 km) and a beautiful coastline with great beaches (3 km). Nearby activities include nature walks, riding, cycling, canoeing, and fishing (trout and salmon). Gear and tackle are for hire at Muriances.

How to get there

By car: N634 from Santander towards Llanes. Exit at Bustio and follow signs to Villanueva and Colombres. Just outside Bustio, turn left at the Agroturismo Muriances sign (on the left-hand side). Then drive about 1 km until you see farm on your right. **Public transport:** From Oviedo or Bilbao, local FEVE train to Unquera. Or Alsa bus to Unquera. Then walk the remaining 2.5 km or call ahead to be picked up.

48. La Valleja

Paula Valero Sáez &
Antonio Rozalén
Aldea Rieña, 33576 Peñamellera
Alta, Asturias
Tel: 985-92 52 36 Tel/fax: 985-41 58 95
valleycas@yahoo.es
Open all year Language: E, GB
€ pn Lodging: 2p 30(ls)-39(hs); breakfast 3;
dinner 10; lunch-pack 6; VAT 7%

Farm and surroundings

This typical Asturian country home, built in 1927, stands in one of the lovely green valleys so typical of the region. The view of the Picos de Europa and the Sierra de Cuera is stunning. The recently restored house with beautiful chestnut support beams has 5 double bedrooms, each with its own bathroom (1 of the rooms is wheelchair accessible). All original building materials were re-used in the renovation. Delicious organic meals are served (vegetarian available). Don't miss the family dinner.

On the farm, Antonio and Paula grow organic fruit. Antonio is also working on a permaculture experiment. Jam is produced in a small house further down the valley.

There is excellent swimming in the Cares river, within walking distance from the lodgings. The farm is roughly 5 km from the ocean as the crow flies, but 30 km away by road. There are countless gorgeous beaches on the coast. The Picos de Europa and Sierra de Cuera offer outstanding walks. On a nearby farm, you can see how sheep's cheese is made.

How to get there

By car: 75 km W of Torrelavega. Take San Sebastián-Santander motorway towards Oviedo. In Unquera, N621 to Panes, from there head for Arenas de Cabrales on the AS114. After about 11 km, AS345 to Alles and Ruenes. 3 km after Alles, turn right on to ascending road to hamlet of Rieña. **Public transport:** Train to Torrelavega, then local FEVE train to Unquera. From Unquera, take an Easa bus to Niserias. Call in advance to be picked up.

49. La Montaña Mágica

Carlos Bueno & Pilar Pando
Cuanda, 33508 El Allende de Vibaño (Llanes),
Asturias. Tel: 985-92 51 76 Fax: 985-92 57 80
www.helicon.es/cuanda.htm
Open all year Language: E, GB, F
€ pn Lodging pn: 1p 37(ls)-50(hs); 2p 46-57(ls)
64-72(hs); suite 65(ls)-85(hs); breakfast 4.50;
lunch & dinner 12; VAT 7%

50. El Correntíu

María Luisa Bravo Toraño &
José Luís Valdés García
Sardalla 42, 33560 Ribadesella Asturias
Tel/fax: 985-86 14 36
elcorrentiu@fade.es www.elcorrentiu.com
Open all year Language: E, F, GB
€ *pn Self-catering: 46-65(ls) 55-85(hs);*
VAT 7%

Spain

Farm and surroundings

La Montaña Magica ('the magic mountain') is a fantastic complex of old farmhouses, situated high up in the mountains with an overwhelming view of the Picos de Europa (25 km away). This is a very quiet and remote location. Carlos and Pilar, who have two small children, welcome you to their environmentally friendly paradise.

Twenty-six sheep and 11 endangered Asturcón horses graze around the farm. Guests may ride and groom the smaller horses. In a dale nearby there are several greenhouses where organic produce is grown for the kitchen. Vegetarian meals are available.

The lodgings have beautiful rooms with baths. Two of these (for up to 4 guests each) have an extra sitting room with open fireplace. There is also a common sitting room with fireplace and a dining room where meals are served. Woodcarving and riding lessons are given on the premises. Bicycles can be hired on the farm. Hiking and riding paths from the farm lead to caves with prehistoric drawings and other sights. There are outstanding beaches 10 km away.

How to get there

By car: From San Sebastián (Donostia), E70 or N634 towards Oviedo via Santander. At km marker 307, exit towards Balmori and Celorio, then AS263 to Posada. In Posada, turn left towards Cabrales (AS115). Continue until La Herreria, make immediate right towards El Allende (past a narrow bridge), then right again on to a narrow road. Follow signs to Montaña Mágica. **Public transport:** Train or bus from Bilbao or Oviedo to Llanes. Then local train or bus to Posada de Llanes. Take taxi for the last 8 km.

Farm and surroundings

This restored farm is situated in a traditional Asturian hamlet on the bank of the Sella river, between the mountains and the sea. Wherever you look there are meadows and oak and birch woods. There is a small, fully furnished guest house with a simple, rustic interior and a fireplace. There are also 2 apartments in the converted granaries. Extra beds are available, including a cot and a crib. You can enjoy the home-grown fruits and vegetables free of charge, as well as the home-made cider, jam and liqueur. In the garden you will find all sorts of herbs to cook with. Guests may help on the farm. The owners do their part to protect old Asturian animal breeds by raising Xalda sheep and an Asturcón horse. Children can play safely in the very large garden.

In the town of Ribadesella (2 km away) bicycles, canoes and horses are for hire and there is a swimming pool. Just a stone's throw from the farm, there are several caves: Tito Bustillo, El Cierru, Cova Rosa and Pozu de la Cerezal, where you will discover 17,000-year-old drawings. Retrace the steps taken by dinosaurs 150 million years ago, between the beaches of Playa de Vega and Pedral de Arra. The famous Picos de Europa national park is 30 km away.

How to get there

By car: From Gijón, AS632 to Ribadesella. Just before bridge over the Sella, turn right towards Cuevas and Sardalla. After 2 km, you reach El Correntíu (on your left). From Oviedo, AS634 to Ribadesella, turn left and cross bridge over the Sella, take immediate left towards Cuevas and Sardalla. **Public transport:** From Oviedo, Bilbao or Santander, train (FEVE) or bus (Alsa, Easa or Turytrans) to Ribadesella. Then walk (2 km), taxi, or arrange pick-up.

51. Posada del Valle

Nigel Burch
33549 Collía (Arriondas), Asturias
Tel: 985-84 11 57 Fax: 985-84 15 59
hotel@posadadelvalle.com
www.posadadelvalle.com
Open Apr-Oct 15th Language: E, GB
€ pn Lodging: 2p 50(ls)-59(hs); extra bed 16;
breakfast 5.50; dinner 14; VAT 7%

Hotel and surroundings

This family hotel was originally an Asturian farmhouse occupied by a priest. It is in a rural setting, surrounded by 5 ha. of land with large apple orchards. The eight varieties of apple grown there are used to produce cider. A herd of Xalda sheep graze in the orchards. These endangered indigenous sheep are cross-bred with old Basque Caranzanas. The house is built into the side of a mountain on a large crag. All bedrooms and common areas have a wonderful view of the valley and the impressive mountain range. Produce from the organic vegetable garden is used whenever available. Dinner must be ordered in advance. Tourist information is available at the hotel and your host can

provide you with information about guided and unguided excursions. Guests who smoke are kindly requested to do so outside.

In the Sella river you can swim, canoe and fish. Near the hotel you will find the Picos de Europa national park and the Reserva Nacional del Sueve (3 km away). Riding opportunities and beaches are also in the vicinity.

How to get there

By car: Santander-Oviedo E70/N634 to Arriondas. From city centre, head for Colunga, Mirador del Fito (AS260), after 1 km turn right towards Collía (AS341). Drive through Collía, after 300 m, turn left on to narrow, unpaved road and drive downhill to hotel. **Public transport:** From Oviedo, Bilbao or Santander, train (FEVE) or bus (Easa or Turytrans) to Arriondas. Then walk the 2 km to the hotel or take a taxi.

52. Posada Ecológica L'Ayalga

Conchy de la Iglésia Escalada &
Luis Díaz González
La Pandiella s/n, 33537 Piloña (Infiesto)
Asturias. Tel: 985-92 30 50 Mob: 616-89 76 38
layalga@terrae.net www.terrae.net/layalga
Open all year Language: E, GB
€ pn B&B: 2p 36(ls)-43(hs); extra bed 6;
dinner 6; lunch-pack 4.80; VAT 7%

Farm and surroundings

L'Ayalga is a traditional Asturian farmhouse near the village of Pandiella in central Asturias. The house deserves the name 'ecological inn', as owners Luis and Conchy do their utmost to conserve the environment. The house was renovated using environmentally friendly materials and energy-saving methods (passive solar power). There are 5 double bedrooms with *en suite* bathrooms. Each room has its own ambience and good view; all the furniture and soft furnishings are made of natural materials and fabrics. The water is also managed in an environmentally friendly way; greywater is separated and recycled, rainwater is collected for various uses and the water is heated by solar power. Home-grown produce is used to prepare tasty, healthy meals. If you want to go on a day trip, the owners will provide a picnic basket. They also organise nature walks and workshops in Tai Chi and Chi Kun. You can also learn how to make cleaning agents and cosmetics from natural products. You can read in the library, play board games and even order a massage. Guests are always welcome to help on the farm. One percent of the rental fee goes to a charity of your choice.

The area is great for walking and cycling. The mountainous Parque Natural de Reres 30 km away, is covered with beech and oak forests. Higher up in the mountains, you will find juniper bushes and crowberries. The forests are home to the capercaillie, bear, wolf and chamois. Further north is the Paisaje Protegido Sierra del Sueve where a significant number of Asturcón ponies roam. To the east you will find the famous Parque Nacional de los Picos de Europa.

How to get there

By car: On N634, at km marker 361, turn off to Infiesto. Drive through village. At edge of village, turn left on to AS254 to Campo de Caso. After 3 km, turn left to La Pandiella. **Public transport:** Train or bus to Infiesto. From there, take a taxi or call to be picked up.

53. La Quintana de la Foncalada

Severino García González & Daniela Schmid
Foncalada 18, 33314 Argüero (Villaviciosa), Asturias
Tel/fax: 985-87 63 65 Mob: 609-97 63 25
foncalada@asturcon-museo.com
Open all year Language: E, F, G
€ pn *Lodging:* 2p 30(ls)-42(hs); extra bed 12; *Self-catering:* 60(ls)-78(hs); breakfast 3; *lunch & dinner 15*

Farm and surroundings

This is a typical coastal farm, nestling among others in a rolling agrarian landscape. Just 4 km from beautiful beaches and friendly fishing villages, this is where Severino and Daniela have raised their children. Several guest rooms share one common kitchen/sitting/dining area. Breakfast is served in this spacious room each morning. For more privacy, there is also an apartment with its own kitchen. A library on the premises offers a wealth of regional information, including local walking guides. The owners, artists by trade, teach guests to make pottery modelled on 17th and 18th century earthenware from the Villaviciosa region. They also actively breed two indigenous endangered animal species: Xalda sheep and Asturcón horses. The farm houses a museum about Asturcón ponies and other indigenous breeds, and several riding horses are available on the farm. Guests can learn to make jam and cider (the traditional local drink), or learn about sheep and horse breeding. Both cider and jam are made from fruit grown on the premises. There is a large garden and a swimming pool for children to play in. Eco-tours to working organic farms in the neighbourhood can be arranged. During the off season, people with an interest in ceramics or horses can work on the farm in return for room and board. Just 8 km away, you will find the Ría de Villaviciosa nature reserve, an area with a unique ecosystem where many aquatic birds spend the winter or rest during their migration.

How to get there

By car: San Sebastián-Santander-Villaviciosa towards Gijón (N632/E70). Continue on the N632 1 km past Villaviciosa, across the Ría de Villaviciosa, then turn right on to the AS256. After about 7 km, turn right (between km markers 4 and 5) on to the VV4 towards Argüero. At the first fork (after approx. 1 km), keep right and then follow the signs to Foncalada. **Public transport:** Train or Eurolines bus to Gijón or the train to Ribadesella. Bus (either Alsa or Cabranes) from Villaviciosa or Gijón to the Cuatro Caminos bus stop in Argüero. From there, walk 2 km or call to be picked up.

54. La Casa del Naturalista

Patricia del Valle Varillas
Quintana de Argüerín 24,
33314 Argüero (Villaviciosa),
Asturias. Tel: 985-97 42 18
josearboleya@hotmail.com
Open all year Language: E, D
€ pn Lodging: 2p 30(ls)-42(hs); breakfast 3; lunch & dinner 11; VAT 7%

Farm and surroundings

La Casa del Naturalista is just a stone's throw from the rocky coastline. Patricia lives with her parents Isabel and Luís in this farmhouse, which has been restored to reveal the original stone walls and oak beams. The farm still features two original *horreos*, the traditional rectangular granaries of Asturias. The old stables have been converted into guest lodgings, with sitting and dining rooms at ground level and bedrooms upstairs. The large garden has a nice sitting area with a view of the fields. The farm also has beehives. The greywater produced on the farm runs through a natural purification system and into a pool so clean that it is inhabited by rare amphibians. Not only is Patricia an organic farmer; she is also a nature guide and works as a co-ordinator at an ecology educa-

tion centre. Patricia's parents help her to manage both the farm and the guest accommodation. Her mother Isabel is known to cook a fabulous *fabada asturiana*, a local bean speciality.

There are many walking and cycling paths in the vicinity. Two km from the farm, you will find the PR (short distance route) from Playa de Merón to Tazones, which runs past 12 old mills. The beach is 3 km away, and the Asturcón Horse Museum is 2 km from Quintana de la Foncalada. Just 9 km away you will find the Ría de Villaviciosa nature reserve, a rare ecosystem where many aquatic birds spend the winter, or rest during migration.

How to get there

By car: San Sebastián-Santander-Villaviciosa towards Gijón (N632/E70). Continue on the N632 1 km past Villaviciosa, across the Ría de Villaviciosa, then turn right on to AS256. After about 7 km, turn right (between km markers 4 and 5) on to VV4 towards Argüero. At first fork (after approx. 1 km), keep left towards Argüerín, then follow signs to La Casa del Naturalista. **Public transport:** Train or Eurolines bus to Gijón or train to Ribadesella. Bus (either Alsa or Cabranes) from Villaviciosa or Gijón to Cuatro Caminos bus stop in Argüero. Then walk 3 km.

55. La Llosa de Fombona

María Rosa Sánchez Martínez
Santa Eulalia de Nembro, 33449
Fombona (Luanco), Asturias
Tel: 985-97 56 46 Mob: 617-78 58 38
Open all year Language: E, GB
Lodging pn: 2p EE72.12(ls)-EE84.14(hs); breakfast 6; VAT 7%

Farm and surroundings

Hilly farmland, cattle and crops - this is the setting for this scenic farm, just 3 km from the beautiful beaches and fishing villages along the coast. The owners were among the first organic farmers in Asturias. They make their living from vegetable growing. In 1998, María Rosa decided to convert part of the farm into guest lodgings. There are four comfortable rooms with stylish decor, each with a bathroom and a television (which can be removed on request). The roomy sitting/dining room has a lirary and an open fireplace. All furniture has either been passed down in the family or bought second-hand and restored by María Rosa herself. The house is filled with the wonderful aroma of the oil used to maintain the woodwork. Fresh milk comes from the neighbour's dairy cows, and bread is baked daily on the premises. There are chickens, which provide fresh eggs, as well as ducks, rabbits, a donkey and two horses of the Asturcón breed.

There are footpaths along the rocky coastline (the Cabo Peñas protected landscape), including a route leading to the Cabo Peñas lighthouse. There are several excellent beaches and pretty fishing villages (Luanco, Candás).

How to get there

By car: San Sebastián-Santander-Gijón, towards Avilés. On ring road at Gijón, exit for Tabaza en Candás (AS19). After 4 km turn right towards Candás (AS239), then go to Luanco. In Luanco, briefly take AS238 towards Avilés, then turn right towards Bañugues until sign to Fombona. At that crossing, turn left and left again after 50 m, then look for sign. **Public transport:** Train or Eurolines bus to Gijón, then Alsa bus directly to Luanco (or FEVE train to Candás then bus to Luanco). From Luanco, taxi or walk (3 km).

56. Camping Fragadeume

Olga Balado Gonzalez
O Redondo s/n, 15617 Monfero, A Coruña
Tel: 981-19 51 30 Mob: 626-76 33 96
rieume@jazzfree.com
www.campingfragadeume.com
Open Mar-Oct Language: E, Ga, GB F
€ pn Camping: tent 2.50-3; caravan 3.50; camper van 6; adult 3; Self-catering: 22-35; breakfast & lunch-pack 3; lunch & dinner 6

Spain

Campsite and surroundings

Fragadeume campsite, named after the nearby Parque Natural As Fragas do Eume, is a former farm. The stables and granary have been converted into a cosy restaurant/bar and common room. There is a large grassy field to pitch your tent on, and there are log cabins for rent. Children can help tend the farm animals, swim in the pool and enjoy the playground. Olga and Juan Carlos teach bread baking and cheese making, and organise excursions to the nature reserve.

Several footpaths cross at the camping ground. One of them leads you through the river Eume gorge and along the reservoir (3 km), while another is a long distance path between the monasteries of Caaveiro and Monfero. The Atlantic Ocean is just 15 km away, and you can take day trips to Ferrol, A Coruña, and the famous pilgrimage site of Santiago de Compostela. The nature reserve - including the prettiest area around the Caaveiro monastery - is one of the last remaining Atlantic coastal forests in Europe. The indigenous trees you will find here include oak, ash, chestnut, willow, *madroño* (strawberry tree), maple, hazelnut and laurel. Along the river banks are various ferns which have been growing since the Tertiary era.

How to get there

By car: From Pontedeume (between A Coruña and Ferrol), take the AC150, then AC151 to Monfero. At Ponte da Pedra, follow signs to the camping ground. **Public transport:** Take a train to Pontedeume. From there, take a Monfero bus and get off at the Ponte da Pedra bus stop. From there, take a taxi or call to be picked up.

57. Casa Pousadoira

Begonia de Bernardo Miño
Lugar de Pousadoira 4, 15635 Callobre
(Miño), A Coruña
Tel/fax: 981-19 51 18 Mob: 629-28 05 65
Open all year Language: E, GB
€ pn B&B: 2p 38.95(ls)-48.68(hs); extra bed 12.02; dinner 9.02; VAT 7%

Farm and surroundings

This former farm stands at the edge of a small hamlet, 9 km inland from the north-western coast of Galicia. It belongs to Begonia and her husband, an enthusiastic young couple who began growing grapes in the year 2000. They also have chickens and rabbits on the property. The property features a *horreo* (a traditional granary on stilts). Guests have a view of the crop fields and woods.

The house has been renovated and decorated with lots of character. It has five double rooms with bath, and common sitting and dining rooms. Breakfast is included in the room price, lunch and dinner are available. All meals are prepared mainly with produce from the organic vegetable garden on the premises.

The surroundings are hilly and wooded. There are beautiful footpaths for walks in the mountains, twenty km further inland. The Lambre river, offering swimming and fishing, is only 500 m from the house. The town of Miño has a beautiful beach and offers many recreational activities on the water. It is also one of the towns on the Camino de Santiago trail.

How to get there

By car: From Lugo go towards La Coruña, take N651 towards Ferrol, take Miño exit. In Miño, follow signs to Villamateo. Continue for 5 km to intersection with bar on corner. Turn right, towards Casa Pousadoira (another 4 km). **Public transport:** Train to A Coruña, then train or bus to Miño. From there, take a taxi.

58. Casa Paradela

Manuel Rodríguez & Paz Domínguez
Carretera Barrio Paradela
32780 Pobra de Trives, Ourense
Tel/fax: 988-33 07 14
1pa712e1@infonegocio.com
Open Jan-Dec 24th Language: E, GB
€ pn Lodging: 47.80; breakfast 5.15; VAT 7%

Farm and surroundings

This imposing 16th century farmstead is surrounded by a hilly landscape dotted with other farms. The farm's numerous sheep graze in the nearby fields. In front of the farm is a small vegetable plot. The farm building is a historical monument with its own chapel, a *sequeiro* (an area for drying chestnuts) and a traditional Galician granary. In 1994, the farm was renovated with meticulous attention to historical and stylistic detail. It has five deluxe guest rooms with bath. Breakfast is served in the dining/sitting room. The surrounding area is ideal for walking and cycling (bicycles can be borrowed at the farm). In winter, guests can make use of nearby ski slopes. Bikes to rent.

How to get there

By car: From Ponferrada N120 towards Ourense and A Rúa. Past A Rúa, take the 536 to Puebla de Trives. Drive through village. Outside village, go over the bridge and turn right at the sign to Casa Paradela. Drive another 2 km, the farm is on the right. **Public transport:** ask for details when booking accommodation.

Capra Hispánica, Sierra de Gredos, Ávila, Castilla y León

Castilla y León, Madrid and Castilla-La Mancha

Castilla y León

The autonomous region of Castilla y León covers an area of about 100,000 km², almost one-fifth of Spain. It comprises nine provinces: Valladolid, Soria, Segovia, Burgos, Ávila, Zamora, Salamanca, León and Palencia. The region's natural heritage dates back millions of years, and it has been inhabited for thousands of years old. The result is a region rich in culture and varied in landscape. Some types of landscape cover hundreds of square kilometres, such as the Sanabria y El Bierzo, the Cordillera Cantábrica, the Sistema Ibérico, the Sistema Central and the Meseta del Duero. The people of this region are well aware of their special heritage, and strive to sustain and restore it so as to pass it on to future generations.

Today's landscape results from aeons of gradual development, and also sudden upheavals. Rivers, lakes, river deltas, beaches and mountains were formed, subsumed and reformed, finally leaving the area with a nearly equal division between mountains and plains. Looking down on the area from space, it resembles a walled castle: towering mountains of 2,500 m surround a large interior plain. Mountains form barriers, which are levelled in places by the rivers that cut through them. The Duero river is Castilla y León's main artery, and the gateway to Portugal (where it becomes the Rio Douro). On its way to the Atlantic, it collects water from the many tributaries that come down from the snow-capped peaks.

There is a wealth of scenery, thanks to the granite and limestone rock formations, the colours of minerals and plants, sharp peaks hiding crystal-clear glacial lakes and deep dark evergreen and deciduous forests. This is bear and wolf country, but the area's wildlife also includes chamois, mountain goats, deer, and various birds of prey. Castilla-Léon has many national and regional parks that preserve this rich biodiversity: La Montaña de Covadonga, the Picos de Europa, Las Batueces and Los Ancares in León, the Gredos in Ávila, Sanabria Park in Zamora, Fuentes Carrionas y del Cobre in Palencia, and the Sierras de la Demanda y de Urbión in Burgos and Soria.

The plains are covered by grass and yellow wheat, and clumps of dark evergreen brush. The horizon is broken now and then by a river rushing through a deep, verdant gorge teeming with wildlife. Magnificent examples are the Alto del Sil in León, the Ebro in Burgos, the Arribes between Zamora and Salamanca, Las Hoces del Duratón y del Riaza in Segovia and the stunning Cañón del Rio Lobos on the Soria-Burgos border. Many of these landscapes seem untouched by humanity. In the *karst* area of Ojo de Guareña in Burgos, for instance, the air is pure and the peace and quiet are almost mystical. Another type of magic can be experienced in deserted ancient settlements, such as the Roman mines at Las Médulas in León.

The first settlers in the region lived in small hamlets in the Cantabrian Mountains, as far south as the Sierra de Guadarrama. Archaeological finds include a Stone Age settlement at Atapuerca in Burgos, where the oldest traces of human habitation in all of Europe were found. More recent, but still very old, are the dolmen at Burgos and Salamanca, the Celtic core of towns such as Ávila and Soria, the Roman cities of Numancia, Clunia and Astorga and a pre-Roman type of artwork unique to Europe.

The region also abounds with traces from the Middle Ages. At that time, Christians, Jews and Muslims populated the nation and left behind a series of buildings, including castles and city walls with Romanesque and Gothic bell towers. Examples of these architectural styles can be found in Segóvia, Ávila and Salamanca, towns which have recently been added to UNESCO's World Heritage list. The region's rich past also includes *pantheons*, warriors, saints, queens, the legend of El Cid, the cross and sword of Emperor Carlos, palaces, monasteries and cathedrals, and last but not least, the Camino de Santiago.

By Fdo. Vidal Postigo Escribano
Historian, Castilla-León Regional Guide

Madrid

The Madrid region is home to both the magnificent capital of Spain and great natural wealth. The many marked walking paths in the mountains north of Spain's central plateau lead you past rivers and reservoirs where aquatic birds spend the winter. There are matchless views over the mountains, with grand castles and picturesque villages. Some 50 km NW of the capital you will find El Escorial, an impressive 16th century palace boasting a stunningly beautiful library and countless artworks. The palace is in the foothills of the Sierra de Guadarrama, a mountain range that is part of a regional park. In the same area, you will find General Franco's oppressive monument to the victims of the Civil War - which was carved out over the course of 16 years by Republican prisoners of war.

Castilla-La Mancha

Castilla-La Mancha is characterized by desolate plains, rolling hills, woods and impressive mountain ranges. You will recognize the windmills on the plains and the many castles from Cervantes' tales of Don Quixote.

Besides its profound historical heritage, this region has great natural beauty. Tablas de Damiel national park is excellent for bird watchers, and Alcalá del Júcar is famous for its jagged limestone cliffs. There are densely wooded areas throughout the five provinces, and there are wine-growing regions everywhere: La Mancha has a very large area of vineyards. Other crops grown here include saffron and olives for olive oil.

In the northern province of Guadalajara you can expect vistas of incredible gorges, spectacular mountains and lakes, an unrivalled wealth of flora and fauna, and foothills where the vines grow in red soil.

La Garganta del Cares, Caín, León, Castilla y León

59. Camping Brejeo

*María del Carmen do Nacimiento &
Eliseo García González*
C/ Saavedra s/n, 24516 Vilela, León
Tel: 987-54 20 25 brejeo@mixmail.com
www.campingbrejeo.es.fm
Open all year Language: E, Ga, GB, PT
€ *pn Camping: tent 2.10-2.70; caravan 2.70;
camper van 3.01; adult 2.70; VAT 7%*

Eliseo and his wife Maria run an ecological campsite called El Brejeo on a quiet spot on the fringe of the village of Vilela, close to Vilafranca del Bierzo. They grow fruits, nuts and a variety of vegetables and own some hens and a horse.

Though quite spacious, the campsite is very quiet with a natural pond and a group of ancient chestnut trees. The main building has a solar powered heating system for hot water supply. Facilities are clean and modern. There is a bar where the locals meet in the evenings and at weekends. In the beginning of August, an important local musical event takes place with Celtic music; in November they have big celebrations reaping and roasting chestnuts. They plan to open their vegetable garden and fruit orchard to their guests, so you will be able to pick your own organic food for a small fee.

This part of Castilla and León is barely known to travellers, and has some truly virgin natural beauty on offer, with an abundance of wildlife and diversity of plants and wild flowers. Numerous long and short walks are available and in the nearby mountains lie the desolate village of Campo del Agua, where large remains of a comet have been found, and the village of Santiago de Panelba, with its 11th century church and panoramic views of the surrounding mountain ranges. The awkward shapes and colourful rock formations of the Roman Mines called Las Medulas are a short drive away, as are the city of Astorga with its cathedral, and palace designed by Gaudí.

How to get there
Villafranca del Bierzo lies 13 km W of Ponferrada.The campsite is not signposted from the A6 or the NV1, but coming from the village of Villafranca it is well indicated.

60. Albergue Las Amayuelas

Cristina Ortega
Plaza de la Iglesia 9, 34429
Amayuelas de Abajo, Palencia
Tel: 979-15 41 61 Fax: 979-15 40 22
amayuelas@cdrtcampos.es
www.cdtrcampos.es/pijtc/municipi.htm
Open all year Language: E, GB, F
€ *pp Groups: B&B 9; full board 19;
lunch & dinner 7; VAT 7%*

Hostel and surroundings
In 1980, Las Amayuelas de Abajo was practically a ghost town. Only one family still lived there; everyone else had left for the city. In 1990, a group of young people decided to breathe new life into the village and its farms and to run them with an emphasis on environmentally friendly principles. There are now four families living there, and several ecological projects are in progress. Thirteen ha of land are used for growing grains and vegetables (all organic). There is an organic sheep and chicken farm where indigenous Black Castillian chickens are raised in order to preserve the breed. The town also has a bakery with a traditional wood oven, an organic store, a bar, a small construction company that builds by traditional techniques, and a hostel. The hostel has five rooms, for 2, 4, 8, 12 and 16 guests. Bathrooms and toilets are shared. There is a sitting room, dining hall, library and two halls for large gatherings. Meals are prepared from locally grown ingredients. Guests can borrow bicycles to tour the surrounding area. Eight km away, you will find the Camino de Santiago trail. Just 1 km from the hostel is the Canal de Castilla, and there are also Roman ruins nearby.

How to get there

By car: From Palencia, N611 north. After approx. 18 km, near Amusco, P983 west towards San Cebrián de Campos and Calahorra de Ribas. After 4 km, turn N towards Las Amayuelas de Arriba and Abajo (2 more km). **Public transport:** Train or bus to Palencia, then train to Amusco. Then walk 4 km. Groups may ask to be picked up. Groups larger than 10 can get 60% reduction on cost of train tickets.

61. El Linar

Carlos de Tomás
37659 San Martín del Castañar, Salamanca
Tel: 923-16 11 16 / 932-43 72 01
Fax: 923-43 75 66
Open all year Language: E, GB
€ pn B&B: 2p from 55; breakfast 4;
dinner 11; VAT 7%

Hotel and surroundings

The small hotel is situated in the stunning, unspoilt and densely wooded Sierra de Francia mountains, 90 km south of Salamanca. This modern, roomy and light structure, has attractive rooms, each with private bathroom. There is a good restaurant (also caters to vegetarians), a bar, a sitting room with open fireplace, a common room and a sunny terrace with a breathtaking view of the Peña de Francia. The hotel borders on a nature park (4 ha), designed and owned by El Linar. The (indigenous) chestnut and oak forests have been partly reforested, and the park is run with an emphasis on nature conservation and education.

There is a small nature museum. Maps showing walking paths in the vicinity are available. The waiter doubles as a mountain guide. This is

also an ideal cycling area. There are no fewer than five medieval villages nearby, all cultural heritage sites.

How to get there

By car: From Salamanca, take the C512 to Vecinos. Then the SA210 to Tamames. 2 km past Tamames, take the C525 S to San Miguel del Robledo. From this village, follow signs to San Martín del Castañar. You will soon see the hotel on the right hand side. **Public transport:** Ask for details when booking.

62. El Burro Blanco

Paul Zegveld & Govert Dibbets &
Yvonne Arends
Camino de las Norias s/n
37660 Miranda del Castañar, Salamanca
Tel/fax: 923-16 11 00
elbb@infonegocio.com www.elbb.net
Open Apr-Oct Language: E, GB, D, NL, F
Camping pn: tent & caravan €9; adult €4

Campsite and surroundings

El Burro Blanco is situated in an oak forest which belonged to the Spanish national forestry commission. Since the campsite first opened for business in 1997, biodiversity has increased due to better forest management and improved environmental practices among campers. This simple but pleasant campsite has good sanitary facilities that have a minimal impact on the ecosystem. The site and its surroundings have become part of the Las Batuecas nature park. In it you will find a great diversity of flora and fauna. This is a paradise for both amateur and professional birdwatchers, butterfly and reptile enthusiasts and botanists. The camping ground has its own library.

From the campsite you can take a footpath through the woods or to the uniquely situated medieval village. Swimming is possible at several spots in the river. Dozens of interesting mountain villages can be found nearby.

How to get there

By car: Miranda del Castañar is situated on the C512 (Salamanca-Coria) and 5 km off the C515 (Béjar-Ciudad Rodrigo). There are signs to both the village and the campsite on both of these roads. **Public transport:** Train or bus to Salamanca, then the regular Salamanca-Coria bus (runs twice daily), get off at Miranda del Castañar.

63. La Lobera

Luz Domínguez Galache
Apartado 4, 05400 Arenas de San Pedro,
Ávila. Tel: 920-37 14 13 Mob: 670-64 35 15
Fax: 920-37 20 17 e.repiso@wanadoo.es
www.laloberadegredos.com
Open all year Language: E, GB
€ *pn B&B: 1p 33.17(ls)-40.19(hs);*
2p 46(ls)-50.64(hs); lunch & dinner 9.63; B!

Guest house and surroundings

La Lobera is situated in one of the most beautiful areas of the Sierra de Gredos, at the edge of a regional park. The guest houses have single and double rooms with bathroom. The lodgings are new, but built from traditional wood and stone. The 1 ha property is a peaceful paradise, with water and rock gardens as well as very old chestnut and cork-oak trees. There is a natural swimming pool filtered by plants and

pebbles. The accommodation uses solar energy. Ten percent of La Lobera's profit goes towards a reforestation project. Guests can attend various courses (painting, dance or yoga) or just go their own way. The centre can also be hired by groups who wish to give their own courses. The delicious local dishes served at La Lobera are made with organic ingredients whenever possible. On request, the hosts organize walks past crystal clear waterfalls and gigantic rock formations in the surrounding countryside - chances are that you will come across a Spanish mountain goat, a golden eagle or an otter. In the vicinity, there is Arenas de San Pedro, a 14th century village with gothic and neo-classical architecture. Talavera de la Reina, which is well worth visiting to see the variety of local pottery, is just 45 km away.

How to get there

By car: Take the NV from Madrid to Talavera de la Reina. From there, drive N to Arenas de San Pedro on the N502. From Arenas, take the AV924 towards Guisando, follow signs to La Lobera. **Public transport:** Train from Madrid to Talavera de la Reina, then the bus to Arenas de San Pedro. From there, take a taxi or walk 2 km towards Guisando, follow signs to La Lobera.

64. Albergues y Apartamentos Rurales Calumet

Hector Martínez López
C/ Reguera 25, 28194 Berzosa del Lozoya,
Madrid. Tel/fax: 918-68 70 63
calumet@sierranorte.net
www.sierranorte.net/calumet
Open all year Language: E, GB
€ *pn Self-catering: 90.15-114.19; Hostel*
weekend: 2p 38; 4p 76; breakfast 2; lunch &
dinner 6.01; lunch-pack 4.21; B!

Lodging and surroundings

Berzosa del Lozoya (elev. 1,100 m) is a tiny, quiet village with 150 inhabitants in the foothills of the Sierra de la Mujer Muerta, less than 85 km from Madrid. The village houses many young artists and artisans, whose studios are usually open to the public. The village stages traditional festivals, exhibitions and theatre performances all year long. Calumet consists of

5 self-catering apartments for 4-6 people, and a youth hostel with double and quadruple bedrooms and dormitories for 10 and 20 people. There is a library and a lounge. In the dining room, you are served breakfast, lunch and dinner - all made from local products. Smoking is not permitted inside.

Berzosa has a centre for alternative therapies where vegan meals are served. Three marked walking routes start from the village. The nearby Atazar reservoir offers swimming, canoeing, windsurfing and sailing. Calumet also provides bicycle routes (there are mountain bikes for hire) and organizes rides on horseback. Thrill seekers can go mountaineering and parapenting (similar to hang-gliding) from El Picozo.

How to get there

By car: On N1 Madrid - Burgos, take exit 60 to El Berrueco (M127). Then go N toward Manjirón. After 10 km cross the Presa de El Villar dam. Continue on M127 for another 4 km, then turn left towards Robledillo de la Jara (still M127) and Berzosa del Lozoya. **Public transport:** In Madrid, Continental Auto bus from Plaza Castilla, destination Buitrago (3x daily). Get off in Berzosa del Lozoya.

65. El Descansillo

Carmen Briongos & Chema Iturrioz
Camino de Valhermoso s/n; 19390 Escalera; Guadalajara
Tel: 949-83 12 52 Mob: 608-51 75 44
eldescansillo@eldescansillo.net
www.eldescansillo.net
Open all year Language: E, GB, F
€ pn B&B: 2p 41; half board 2p 62; full board 76; VAT 7%

Hotel and surroundings

In the winter, only 10 people live in this small village - the population swells to 40 in the summer. Carmen and Chema's *hostal* is on the outskirts of the village, with a view of the rolling fields and woods. Chickens, dogs and cats live on the premises. The family, with three children (born in '86, '90 and '94), live on the top floor of this stylish hotel, built by hand using ecologically friendly materials. When cooking for guests, Carmen uses ingredients from her organic vegetable garden. Guests have access to a nicely decorated sitting room and a separate dining room. There are seven double rooms with bath and an attractive terrace. The entire house seems to live and breathe an ecological lifestyle. Carmen makes jam, bakes bread and weaves blankets from naturally dyed wool. Chema's brother is a beekeeper, and makes candles from beeswax. You are welcome to watch him work. The hotel is situated near the Alto Tajo nature park, the upper course of the Rio Tajo. You can take walks to a spectacular gorge. There is also a clear river nearby for swimming and canoeing. Horses can be hired near the hotel.

How to get there

By car: Take the NII Madrid-Zaragoza until exit 135. Take the N211 towards Molina de Aragón. Four km before Molina de Aragón, take the CM2015 to Corduente. Once there, head towards Torete. After 10 km, turn left towards Fuembellida, After 7 km you reach Escalera. The hotel is the last house on the edge of the village. **Public transport:** Train from Zaragoza or Teruel to Monreal del Campo, from there take the bus to Molina de Aragón. Then take a taxi (another 25 km).

Valencia and Islas Baleares

The autonomous region of Valencia (Comunidad Valenciana) is best known for its countless coastal holiday resorts. Inland, however, the atmosphere is far more traditionally Spanish. This fertile region is especially well-known for its orange groves and many vegetable crops. A number of nature reserves are found along the coast. The Parc Naural de l'Albufera just south of the city of Valencia is a fresh water lagoon where some 250 bird species live. North of Benidorm you can find the limestone cliffs of El Peñón de Ifach, and south of Alicante, the Santa Pola salt pans.

Castillo Bellver, Palma de Mallorca, Islas Baleares

El Maestrat, a plateau bordering on Teruel province, is an unspoilt area with picturesque villages. The northernmost corner of Castelló province forms the gateway to Puertos de Beceite, a 30,000 ha nature reserve stretching into Tarragona and Teruel provinces. The park features rough mountain terrain with deep gorges, fields and the biggest beech trees in all of Europe. It is also the habitat of large numbers of chamois, wild boar and partridges. Just across the Castelló-Teruel border, near Valdelinares, is a small ski resort.

The Carrascal de La Font Roja nature park, in the mountains of northern Alicante province, near Alcoy, is known for its lush vegetation. It is the only well-preserved Mediterranean forest on Tertiary limestone, and has a great variety of vegetation and wildlife. The mountains between Alcoy and the coast are extremely suitable for walking: the paths are well-marked and route descriptions are available. The Alicante coast, the Costa Blanca, owes its name to the white and pink almond blossoms that bloom in the mountains every February. Alcoy is famous for its *fiestas de Los Moros y Cristianos*, colourful processions that commemorate the 1492 Christian conquest of the Arabs.

In the south-west, Alicante is bordered by the province of Murcia. Here you will find the Sierra de Espuña and the 13,855 ha nature reserve of the same name. Its large expanses of evergreen forest are home to the moufflon, wild boar, fox, genet, eagles and other birds.

Islas Baleares: Mallorca and Ibiza

Mallorca is the largest of the Baleares, an archipelago of five major islands and several smaller ones, some 240 km east of Valencia. Mallorca's capital Palma is one of the largest ports on the Mediterranean. The city also has many interesting monuments, such as a Gothic cathedral. Owing to its pleasant climate and many beaches, Mallorca has become a major tourist destination. In the interior, villages perch on the steep mountain cliffs; the plains and hills are used for growing grains, fruit and wine grapes. The island has several caves worth visiting, and its western edge is dominated by the Sierra de Tramuntana (1,445 m). This ridge rises steeply from the sea, leaving room only for a few barely accessible bays. This area is perfect for walking, particularly in spring when most trees and plants are in bloom. The northern Parc Natural de s'Albufera is a 2,500 ha nature reserve featuring reed-covered wetlands and dunes, a haven for the Mediterranean sea turtle and other rare species. Eighteen km from Mallorca's southern coast is the little desert island of La Cabrera. This is one of Spain's main nature reserves, accessible only with a special permit or on an organized group excursion from Palma or Colonia de Sant Jordi. One of the main attractions is La Cabrera's rich underwater life. There is a small

Sunset at La Albufera, Valencia

museum and a visitors' centre; wildlife experts take visitors on guided tours which begin at the small harbour.

Ibiza is most famous for its mass tourism and wild nightlife. But there are plenty of other ways to enjoy the island and its natural beauty. Ibiza makes for great walking: the northern side of the island is particularly quiet. The landscape is mountainous, but the peaks are not high; Es Fornás, the tallest mountain, is only 450 m at its peak. Pine-covered hills rise above fertile valleys with olive, almond and fig orchards and the occasional vineyard. You will also see little white churches like the one in Sant Miquel, which are typical of the island. Villages like Sant Joan and Santa Agnés still breathe Ibiza's original, peaceful atmosphere. The island has a very mild climate and rain is unusual.

66. Mas d'Oncell & Mas del Rey

Carmen Ortí Caballer & Enrique Montañés
C/ Sant Joan 66, 12140 Catí, Castellón
Tel: 964-42 83 17 Mob: 639-57 81 12
Open all year Language: E, GB, FR
Lodging pp: €13.22; meals by arrangement

Farmhouses and surroundings

Both Mas d'Oncell and Mas del Rey are 16th century farmhouses, peacefully set in the natural surroundings of Alt Maestrat, in the heart of rural Valencia, 7 km from the village of Catí. Alt Maestrat is rich in historical, geological and natural treasures, with dramatic rock formations, Moorish ruins and an abundance of wildlife and flora.

The stone-built farmhouses have a rustic sense about them, having been renovated with a careful eye for their authentic character, and set amidst mountainous wilderness and dense oak woods. The owners are very conscious about the environment and plan to install solar panels to replace the oil heating.

Mas d'Oncell is the simpler of the two, with an open fire but without any heating system; its old time atmosphere is genuine. It offers two bathrooms with toilets and hot water, five double bedrooms and a kitchen and a living.

The stylish Mas del Rey, though more comfortable, is also in a rural style. There are five doubles, three bathrooms with a toilet each, a large living room with TV, central heating and a garage for a car.

How to get there

Mas d'Oncell and Mas del Rey are 7 km S of Catí on the CV-128. Details with your booking.

67. Mas de Noguera

12440 Caudiel, Castellón
Tel/fax: 964-14 40 74
masnoguera@inicia.es
Open all year Language: E, GB, F
Hostel pp: half board €22; full board €28; B!

Farm and surroundings

Mas de Noguera has been a pioneer in organic farming and cattle raising for the past 15 years. The co-operative serves as a model farm for sustainable rural development and ecological and environmental education. The farm offers courses to adults, professionals and schoolchildren on subjects including farming, cattle raising, beekeeping, food preparation, reforestation, clean energy production, herbs and water treatment. There are 60 ha of land, 3 windmills and enough solar panels to produce the farm's electricity, hot water and central heating. The greywater produced on the farm is organically treated and reused for irrigation. The farm complex has homes for the owners and their personnel, and a beautiful new hostel (min. 12, max. 60 guests). Guests sleep on wooden bunk beds and share common showers and toilets. The common sitting rooms have fireplaces and there is a large room for gatherings. The dining room is in another building, where meals (vegetarian available) are served at fixed times. The meals are made with the farm's own products, such as yoghurt, milk, eggs and bread.

Mas de Noguera is set in the unspoilt mountains of the Sierra de Cerdaña, 9 km from the village of Caudiel and 55 km from Sagunto and the Mediterranean Sea. The farm is the starting point of several marked walking and cycling routes, and is situated alongside the GR-7.

Spain

How to get there

By car: Drive from Barcelona to Sagunto. There, take the N234 towards Teruel, after about 45 km, take the exit to Caudiel (CV195). Pass Caudiel, stay on the CV195 until km marker 10. Turn left at the sign to Mas de Noguera. The farm is 3 km down this narrow asphalt road. **Public transport:** Train to Caudiel (9 km from farm) or the bus to Jerica (15 km away), then take a taxi.

68. La Surera

Calle Carboneras 5, 12413 Almedíjar, Castellón. Tel: 964-13 74 00
Fax: 964-13 73 56
surera@arrakis.es www.surera.com
Open all year Language: E, Ct, F
€ pp Hostel: B&B 16.83; half board 22.84; full board 28.85; lunch & dinner 7.25

Hostel and surroundings

This hostel, built in traditional Spanish style, is located in a mountain village in the Sierra d' Espadà nature park. The hostel is surrounded by cork-oaks and Mediterranean flora and fauna, with olive and almond trees dotting the fields. The host family lives in the inn and maintains a large organic vegetable plot. They sell cheese, honey, potatoes, olives, baskets, natural perfumes and books. La Surera is also a teaching venue where pupils learn about the environment and adults take cooking lessons. There are 3 quadruple bedrooms and 4 dormitories (2 x 9, 2 x 17 beds: all bunk beds). Guests share all facilities: showers, toilets, a common room and a dining room. This lodging is wheelchair-accessible and offers safe play areas for children. B&B, half board and full board are offered.

Guests must bring their own towels and bed linen. There is a washing machine on the premises. Information about the surrounding countryside is available and there are bicycles for hire. Many activities are nearby: riding, swimming, walking and sightseeing. Guided tours of the area are available if there are enough participants.

How to get there

By car: On the A7 (Castellón - Valencia), exit at Sagunto. Take the N234 in the direction of Teruel. In Segorbe, turn right (CV 200) to Castellnovo and Almedíjar. In Almedíjar, take the first street on the left, next to the school and the pharmacy. **Public transport:** Take a train or a bus to Segorbe, then a taxi (8 km).

69. Permacultura Bétera

Arnaldo Llerena Pinta
Carrer Les Llomes s/n, 46117 Bétera, Valencia
Tel: 961-60 01 78 Mob: 626-13 15 56
arnalben@teleline.es
Open all year Language: E, GB
€ pn Self-catering: 80(ls)-100(hs); 560(ls)-700(hs) pw; extra bed 12; VAT 7%

Farm and surroundings

At just half an hour from Valencia and from endless beaches, this quietly located farmhouse, surrounded by fragrant pine trees and orange orchards, sits in an ideal location to combine total relaxation with culture and activities. Arnaldo and Beatriz run this 3 ha farm on a permacultural basis, growing vegetables, potatoes, flowers, trees, fruits, nuts and olives, and have started a vermiculture project. They keep hens, ducks, geese, bees and rabbits and run courses of permaculture and alternative medicine. Massage and manual treatments are available.

The cottage is fully equipped and has 3 bedrooms, allowing 8 people to stay. It is an ideal place for families with children: there is a playing field around the house and a lovely garden to sit and relax.

The mild Mediterranean climate makes this area ideal for a visit, no matter what the season. You can explore the Sierra de Caldona on foot, bicycle or in a horse-drawn carriage whilst admiring the scenic natural surroundings. A long list of places to visit in this historically rich area, including monasteries, ancient villages and castles. The village of Bétera, 5 km away, offers shops, restaurants and a swimming pool, though you may prefer to sit on one of the sandy beaches, which are some 23 km away, and let a mild sea breeze fondle your hair. Spectacular processions and firework shows take place between 12 and 22 of August, when the Virgin of Ascendance is celebrated.

How to get there

By car: from the A7, N of Valencia take exit 494 (Bétera-Burjassot) and follow direction Bétera. Leave village direction Náquera (CV310). You will cross over two bridges over a couple of gorges. Go left 200 m after the second bridge and follow road (Carrer Les Llomes) to the house. **Public transport:** from Valencia take train on Plaza de España to Bétera. Call for a pick-up.

70. Casa del Río Mijares

Albergue Rural - Centro de Interpretación Ambiental
Apartado 59, Aldea de Mijares,
46360 Buñol, Valencia. Tel: 96-212 73 00
mijares@adev.es http://mijares.adev.es
Open all year Language: E, Ct, GB
Hostel pp: full board €28.85; VAT 7%; B!

Farm and surroundings

This large, early 19th century manor house is in a beautiful mountain valley adjoining a national park. The biggest treasure on the 155 ha property is the *nacimiento*. This 'place of birth' - the source of the Mijares river - has given this land a long history of cultivation. In 1996, the property was bought by a group of people, aiming to restore it to its original glory. They earn a living from organic farming and cattle raising, environmental education, arts and crafts and eco-tourism. They have 50 chickens, 30 rabbits, 13 dogs and a donkey.

The house is soberly but tastefully furnished. It has 2 dormitories with bunk beds (sleeping 6 and 8). Guests share the showers and toilets. There is a dining room, and a common room with a fireplace. The meals (vegetarian available) are prepared with home-grown ingredients when possible. In spring and autumn, groups of schoolchildren work on the farm. For the rest of the year, the hostel is open to other guests. Bring your own towels and sleeping bag.

The surrounding mountains teem with flowering plants and wildlife. The hostel is on the GR-7 trail, but there are many local footpaths as well. Carlos, a biologist and local expert, takes you on walks to discover the local flora and fauna. There are great spots for swimming and playing in the river. Children are welcome to help out on the farm. Arts and crafts courses are occasionally organized. Horse and wagon trips are also possible.

How to get there

By car: On the A3 / E 901 from Valencia, take exit 319 to Buñol, then head for Buñol-Yátova on the CV429. After km marker 17, before a stone bridge, turn on to an asphalt road into the woods. The hostel is 1 km down this road. **Public transport:** Local train from Valencia to Utiel, get out at El Rebollar station. Call in advance to be picked up - or take a 4 hr walk through fantastic natural scenery on the GR-7 S until you reach the hostel. Alternatively, take a bus from Valencia to Yátova, from there take the PRV-148 until you cross the GR-7. Walk N to the hostel.

71. Sierra Natura

Diego Lozano
Ctra. Moixent - Navalón km 11.5,
46810 Enguera, Valencia. Tel: 96-225 30 26
s_natura@terra.es www.sierranatura.com
Open all year Language: E, GB, F, D, Ct, Ga
€ pn Camping: tent 3-3.65; caravan 4;
camper van 5.50; adult 3.65; child 2.74;
Lodging: 1p 23; 2p 40; 3p 57; 4p 65;
Self-catering: 65-130; VAT 7%; B!

Campsite and surroundings
Sierra Natura is a nudist campsite in the sparsely populated and unspoilt hills of Valencia. The owners clearly have a sense of humour - one of the containers in which the campers separate their garbage reads *bañadores*, or bathing suits. The small campsite has preserved the natural contours of the terrain. The large pitches have plenty of privacy. Water and electricity are optional. The campsite is also open to caravans and campers. The swimming pool on the premises is fed by 2 natural waterfalls. There is a sauna, a restaurant, a bar/shop, a common room with cooking facilities, and an organic vegetable garden where campers may harvest produce for free. All the buildings were designed by the owner in an idiosyncratic, curved style with playful lines. The old granary has been converted into a number of self-catering apartments and guest rooms.
  The mountainous surroundings are ideal for

walking and biking. Information about day trips is available on the premises.

How to get there
By car: On the N430 Valencia - Almansa, near Moixent, exit to Navalón (CV589). Pass km marker 11, turn right and follow the signs to the camping ground. **Public transport:** Take a train or bus from Valencia to Moixent. From there, take a taxi.

72. Camping El Teularet

Partida de El Teularet s/n, 46810 Enguera, Valencia. Tel: 962-25 30 24 Fax: 962-25 30 42
teularet@pv.ccoo.es http://teularet.ccoo.es
Open all year Language: E, GB, F
€ pn Camping: tent 2.70; caravan 3.50;
camper van 4.02; adult 4.02; child 3.69;
breakfast 3; lunch & dinner 7.21; VAT 7%

Campsite and surroundings
This beautiful, ecological campsite is situated at an elevation of 780 m in the Sierra de Enguera, one of the most unspoilt areas in the province of Valencia. The silence here is incredible. The nearest village is 11 km away. There are 66 pitches for tents, campers and caravans. The hostel has 18 double rooms, each with its own bathroom. The campsite swimming pool is heated by solar panels in winter. More than 70% of the campsite's power is generated by solar panels and windmills. Runoff and greywater are treated and used for irrigation. Special water-saving shower heads and taps have been installed, and garbage disposal is kept to a minimum. There are experiments in organic farming (including chicken and rabbit raising) on the grounds. One of the buildings houses

the 'nature school', where schoolchildren and adults can attend courses in nature and environmental science, organic farming and animal husbandry.

The mountainous surroundings are excellent for walking and cycling, or for taking excursions to caves with prehistoric paintings. Valencia, Alicante and the Mediterranean Sea are only one hour's drive away.

How to get there

By car: South of Valencia, take the N430 towards Almansa and Albacete. Continue until exit to Moixent (left of the road), here turn right on to CV589 towards Navalon. Halfway to Navalon you find entrance to campsite. **Public transport:** Train or bus from Valencia to Moixent. From there take a taxi or walk (11 km).

73. Mas de La Canaleta

Emilia Moratinos Climent
Partida Solana Alta, 03517 La Serrella (Confrides), Alicante
Tel: 965-97 23 60 Mob: 626-15 00 41
Open all year Language: E, GB, S
B&B pp: €18; full board €36; B!

House and surroundings

This old stone house (elev. 900 m) is situated in a pretty, densely wooded and mountainous area of the unspoilt Sierra de Serrella, 8 km from the village of Confrides. Peace and quiet reign supreme. The house has 2 double and 2 triple bedrooms for rent. There is a large common room with a big wood furnace for cold winter nights. The showers and toilets are outside, and have a wonderful view. There are a few solar panels for electricity, and plenty of

candles. The whole property still needs a lot of work. You are welcome to help out for free room and board. Emilia prepares tasty (vegetarian only) meals using organic produce from her own garden and local growers. She uses traditional methods of pickling olives and drying vegetables and collected herbs. Her cosy eat-in kitchen is full of these. Fresh eggs are available.

The surroundings are great for walking and cycling. You can strike out on your own, with the help of maps and itineraries, or you can hire a local guide. You can also accompany Emilia on her visit to the goat herder to buy cheese. This is a one hour walk each way. The steep terrain is unsuitable for young children.

How to get there

By car: On A7 / E15 Valencia-Alicante, take the Callosa/Altea exit (n° 64). Drive to Callosa (CV 755), then continue on same road to Guadalest/Alcoy. Past Guadalest, take CV70 to Confrides and Alcoy. After Confrides, drive another 4 km towards Alcoy until the El Rincón de Olvido restaurant. Call in advance to be picked up here. **Public transport:** From Benidorm, bus to Confrides. Call in advance to be picked up. Or train to Alcoy and then bus to Confrides.

74. La Higuera

Stijn Lohman
Apartado 112, 03510 Callosa d'En Sarrià, Alicante. Tel: 965-97 23 80 Mob: 649-94 32 91
Open all year Language: NL, E, GB
€ pn Camping: tent 4-6; caravan & camper van 6; adult 5; child 3; caravan to rent 165 pw; Self-catering: 200 pw; breakfast & lunch 4; dinner 10

Campsite and surroundings

La Higuera is a quiet campsite filled with the scents of rosemary and thyme. You can pitch your tent on one of the amphitheatre-like terraces of a medlar orchard so characteristic of this region. Extra hands are welcome to help out with the medlar harvest in April and May. Other trees on the grounds include almond, lemon and carob. There is a small vegetable and herb garden, and chickens range free in the farmyard. The camping ground is open to tents, campers and caravans. La Higuera also

Spain

has cabins for rent (with a gas stove and 12-V power supplied by solar panels). The facilities are basic. Guests share a toilet and shower. Only one of the cabins has its own toilet. Arrange in advance to use the kitchen. On request, Stijn also prepares meals which are shared at outside tables on a lovely spot. Eat with a view of the mountains and the sea while the crickets chirp in the background. Water is scarce in this region and therefore must be used sparingly. The greywater is filtered, purified and used to irrigate the vegetable garden. Guests may help do chores on the farm. This part of Spain has a very mild climate even in winter, which makes it an excellent winter holiday destination.

You can take great walks in the surrounding mountains (itineraries available). It is well worth visiting El Arca de Noé, a sanctuary for wild animals (just 8 km away). The sea is about 20 km away.

How to get there

By car: On the A7/E15 Valencia-Alicante, take exit 64 at Callosa/Altea. Drive to Callosa. From there, take the CV755 to Guadalest/Alcoy. At the El Riu restaurant, just before the bridge over the river, call to be picked up. Or continue on the same road for another 1.9 km until you see a small red arrow on a stone to your left. Take the road uphill (sharp curve). After 600 m turn right at a "Y" crossing. After 100 m, take the unpaved road to the left and downhill to the campsite. **Public transport:** From Benidorm take the bus to Callosa d'En Sarrià and Guadalest. Get out 5 km past Callosa at the El Riu restaurant (call ahead to be picked up) or have the bus driver drop you off 2 km further at 'Las Casas Balaguer" (0.8 km from campsite).

75. Ets Albellons

Juan & Sebastian Vicens
Binibona 07314, Caimari, Mallorca
Tel: 971-87 50 69 Fax: 971-87 51 43
finca@albellons.com www.albellons.com
Open all year Language: E, D, GB
€ *pn B&B: 2p 66(ls)-74(hs); dinner 24.04;*
VAT 7%

House and surroundings

This splendid, luxurious farm is situated in the mountainous NW of the isle of Mallorca, an hour's drive from the capital Palma. Perched high in the rough Sierra Tramuntana, Ets Albellons looks out over the flatlands around the town of Inca. The farm is owned by the Vicens family, generations of whom have been sheep farmers. To supplement their income, Juan and Sebastian Vicens converted their parents' farm into a fantastic mountain hotel.

The building has 8 double bedrooms with bathrooms *en suite*. There are also 3 luxury suites, each with a bedroom, sitting room and a bathroom *en suite* with a whirlpool bath. Two of the luxury suites have a private terrace. The property also has a swimming pool. Juan is reforesting and tending the original 40 ha of holly-oak, carob, almond and olive trees. His parents live 1.5 km down the road in Binibona, a hamlet of 3 farmhouses. Here, they raise some 300 sheep as well as pigs, goats, horses, rabbits, quail and chickens. The animals and the produce from their traditional vegetable garden provide the ingredients for the delicious Mallorcan dishes which mother Francisca serves her guests.

Your hosts are happy to take you to see the animals, garden and mountains. You can walk through a breathtaking mountain setting to the Lluc monastery. Also worth seeing are the nearby caves, with stalactites and stalagmites.

How to get there

By car: Ferry or flight to Palma de Mallorca. On the Via de Cintura (ring road) exit to Inca (PM27). In Inca, head for Alcudia (C713) and pass by Selva. In Caimari, turn right at bank (Banca March), continue to church, there turn left on to narrow country lane. At next church, turn right towards Binibona. There, turn right and immediately left on to unpaved road uphill to Ets Albellons (signs). **Public transport:** Ferry or flight to Palma de Mallorca, from there take the train to Inca and then call a taxi.

76. Son Mayol

Apolonia Vaquer
C/ San Nicolao 12, 07200 Felanitx, Mallorca
Tel: 971-18 35 82 / Tel: 971-58 09 00
Fax: 971-58 09 00
Open all year Language: E, D
Lodging € pp: 30.03(ls)-36.06(hs);
Self-catering € pn: 96.16-108.18; breakfast
6.01; dinner 15.03; VAT 7%; B!

Farm and surroundings

This traditional farm is situated amidst almond and fruit orchards in the flat arable land of South Mallorca. The farm is just 15 km from the beautiful coastline with its many beaches (quite crowded in the summertime). Hostess Apolonia runs the farm together with her daughter. She is an old-fashioned farmer, proud to show you her large vegetable garden and the 50 chickens and two pigs that roam the farmyard. She kills the animals herself and processes the meat into various traditional sausages and *sobreasadas*, the special bell pepper sausages of the Baleares, which she uses in her authentic Mallorcan recipes. Vegetarian cooking is not her speciality. Her home makes a hospitable, if somewhat disorderly, impression - for instance she uses the 12th century chapel as a storage shed. The rest of the building, which dates from the 17th century, has been converted into simple apartments and rooms. The old wine cellar under the house is now used as a sitting/dining room with an open kitchen. Guests are free to harvest whatever they need from the garden. There is a beautiful swimming pool on the premises. This is a tremendously peaceful location.

Felanitx is famous for its black earthenware. The monastery of Sant Salvador is not far, and Mallorca's famous beaches are 15 km away.

How to get there

By car: Ferry or flight to Palma de Mallorca. From Palma PM19 and C717 to Campos. There, PM512 to Felanitx and Manacor. At km marker 8, you see sign to Son Mayol. Turn left here. Drive another 3 km on narrow country roads, following signs. **Public transport:** From Palma de Mallorca, bus to Felanitx. Then taxi (7 km).

77. Es Palmer

Dimas Velasco
Ctra Campos-Colònia Sant Jordi km 6.4
07634 Colònia Sant Jordi, Mallorca
Tel: 971-18 12 65 / Tel: 971-46 61 20
Mob: 629-63 98 12 Fax: 971-46 08 89
reservas@espalmer.com www.espalmer.com
Open all year Language: E, GB, F, D
€ pn B&B: 2p 86.84-97.06(ls);
102.17-114.19(hs); dinner 12.02

House and surroundings

This former farm is situated in the flat arable land of southern Mallorca, 5 km from the coast. It is a green oasis in a countryside dotted with old windmills. The farm's garden has fruit trees and herbs, and there is an organic vegetable plot and a swimming pool. The ten stylishly restored rooms are all on the ground floor. They each have their own bathroom and nearly all have a garden or terrace. Breakfast and dinner (also vegetarian) are served in the artistically decorated dining room. The tasty, nutritious meals are prepared with home-grown ingredients, both fresh and preserved. The jars and bottles of preserves add to the country feel of the kitchen. Bicycles and cars are for hire.

The beautiful, unspoilt beach of Es Trenc is only 5 km away. Even in summer you will find peace and quiet on this 6 km stretch of beach, which borders on protected woods and sand dunes. From Colonia Sant Jordi, organized boat excursions take you to one of Spain's national parks: the island of La Cabrera (museum and visitors centre, guided tours possible). In Colonia Sant Jordi and on the beach at Es Trenc sailing, windsurfing, SCUBA diving and other aquatic sports are available.

How to get there

By car: Ferry or flight to Palma de Mallorca, then drive to Campos (via the PM19, PM602 and C717). In Campos, follow signs for Santanyí. Just outside Campos, turn right on to the PM 604 to Colonia de Sant Jordi. At km marker 6.4, turn right at the sign to Finca Es Palmer. **Public transport:** From Palma de Mallorca, take a bus to Campos, then take a taxi (6 km).

78. Can Marti

Isabelle & Peter Brantschen
07810 San Joan de Labritja, Ibiza
Tel: 971-33 35 00 Fax: 971-33 31 12
globalspirit@jet.es
Open all year Language: D, F, E, I
€ pn Lodging: 2p 27.05(ls)-42.01(hs);
Self-catering: 49.58-81.14(ls) 84.14-138.23(hs);
extra bed 9.02; breakfast 6.01; dinner 15.03

Farm and surroundings

This organic farm is set in wooded hills in the relatively quiet northern part of Ibiza. Owners Isabelle and Peter apply various environmentally friendly techniques on their farm, such as permaculture, solar energy and waste recycling. There is 1 double bedroom with a bathroom *en suite* and its own terrace. There are also 3 self-catering apartments (own bathroom, terrace and kitchenette) large enough for 2 adults and 2 children. All lodgings have a great view. No smoking or pets allowed.

The old, thick walls keep the house pleasantly cool even in the summer heat. Children are more than welcome here, and they will be delighted by the chickens, ducks, goats and mule. On request, guests may help out in the environmentally friendly fruit and vegetable garden. There is a Renault 4 for hire. Use of bicycles is included in the price of lodging. Beautiful footpaths, great beaches and a yoga centre are near by. The village centre is a 10 minute walk away. Every Saturday there is an open air market where you can buy organic produce.

How to get there

By car: Ferry from Barcelona or Denia to Ibiza (Sant Antoni Abat). Farm owners will provide itinerary upon request. **Public transport:** Ferry from Barcelona or Denia to Ibiza (Sant Antoni Abat). From there take a taxi or call in advance to be picked up (2,500 pesetas). Or take a bus from the harbour or the airport to San Joan de Labritja.

Playa Torrent de Pareis, at Escorca, Mallorca, Islas Baleares

Extremadura

This is one of the most beautiful - yet least known - parts of inland Spain. The summers are very hot and dry; spring and autumn are the best times for a visit. There are great cities such as Mérida, Trujillo and Cáceres, where the Roman and Moorish civilizations have left their mark. The area also has famous nature parks such as the Parque Natural de Monfragüe in the province of Cáceres (between Trujillo and Plasencia) and the smaller Parque Natural de Cornelio, 18 km NE of Mérida. The mountains of Monfragüe can be seen from afar, but hidden between the cliffs and wooded slopes is a deep valley. This is where the Rio Tao and the Rio Titter merge. Two parallel mountain ridges have both sunny and shady areas that offer refuge to a variety of vegetation and wildlife. In sunny, dry places you can find wild olives and wonderful smelling rosemary and lavender. The damp, shady areas are heavily wooded with oak and cork-oak trees. This is where lynxes lives, and where the rare black stork, the protected golden eagle and other large birds of prey breed. You can choose from five splendid walking paths.

The Parque Natural de Cornelio owes its name to the Cornelio reservoir, which dates from Roman times. This is a protected bird sanctuary with attractive oaks, scrub and green meadows, rivers, waterfalls, granite rock formations and caves. There are two entrances to the park: in the west, via Mirandilla, and from the south, through the village of Campomanes (past Trujillanos). There are some driveable roads in the park, but cycling and walking are recommended.

Most ECEAT places to stay are situated in the Sierra de Gata, a mountain range with peaks from 1,200 to 1,500 metres. This area, in the northern Extremadura near the Portuguese border, is a forgotten corner with mediaeval villages, holly-oak woods, clear streams, unique plants and animals, and stork nests on every church tower. Horse-drawn ploughs and wagons are still a common sight here. This part of Spain also has a rich gastronomic tradition.

Cortijo in Extremadura

79. Finca la Casería

María Cruz Barona Hernández & David Pink
Carretera N110 km 387.5, 10613
Navaconcejo, Cáceres. Tel/fax: 927-17 31 41
mariaydavep@navegalia.com
Open all year except Aug & Christmas
Language: E, GB
€ *pn B&B: 2p 40-60; Self-catering: 60-84;*
375-540 pw; dinner 16.82

Farm and surroundings

This 50 ha farm, built on the ruins of an old
Franciscan monastery, is surrounded by woods
and meadows. Its remote location in the middle of the Jerte valley offers a splendid view of
the surrounding mountains. Proprietors María
Cruz and David make a living by raising cattle
and growing cherries and plums. The living
room is furnished with antiques, and the dining room has an open fireplace. There are six
bedrooms (four with bath), a common sitting
room and three apartments: *El Palomar* (sleeps
2 and a child), *La Casita* (sleeps 2) and *El Silo*
(sleeps 4). A swimming pool brings refreshment
in the summer. A crib and a washing machine
are available. The kitchen may be used for an
additional fee. The accommodation is suitable
for the elderly, and for guests with minor dis-
abilities. Breakfast is included in the room
price; evening meals (available for groups only)
are made mainly with organic fruit and vegetables that are in season. Some meals are suitable
for vegetarians. This is an environment-friendly
household (with ecological water purification,
for instance). Guests are welcome to help out
on the farm.

The surrounding area is excellent for cycling,
walking, riding, swimming (waterfall at 1.5 km)
and other water recreation. There are muse-

ums in Plasencia (30 km away). The surrounding countryside offers churches, a hermitage, a
nature park and panoramic views. The owners
organize all-day mountain excursions, and can
provide information about the local birds,
plants, flowers and medicinal herbs.

How to get there

By car: Navaconcejo is on the N110, about 30
km NW of Plasencia. In the village, ask for directions to Finca la Casería (another 3 km
away). **Public transport:** Train or bus to Plasencia or Avila. From there, take the Plasencia-Avi-
la bus line, get out at Navaconcejo. Ask for directions to Finca la Casería (3 km away).

80. Finca Las Albercas · El Becerril

Carlos Donoso &
Carmen Lobete Ondarza
10857 Acebo, Cáceres
Tel: 927-14 17 24 Mob: 689-40 07 50
ecogat@inicia.es
Open all year Language: E, P, F
€ *pn Self-catering: 43-150(ls) 51-210.50(hs);*
VAT 7%

Farm and surroundings

At the end of a country road, 1.5 km from the
village of Acebo, you will find the Las Albercas-
El Becerril farm, surrounded by pristine mountains and indigenous chestnut and oak forest.
Owners Carmen and Carlos make a living by
selling their organically grown kiwis, apples,
oranges, grapes, olives, almonds and other
nuts.

Carlos and Carmen grow their produce on a
small plot, and guests are welcome to help out.
They have set up a local NGO to support organ-

ic agriculture and cultivate indigenous plant varieties, and offer courses in organic farming, sustainable energy, cheese-making and bee-keeping. Two cows provide fresh milk, and chickens deliver the eggs. The surroundings are excellent for mountain walks. The farm has bicycles for hire, and donkeys for those who prefer to ride. Guests can join Carlos on excursions to discover the medicinal herbs growing in the vicinity. The mountain stream running alongside the farm offers clean drinking water. There is also a natural swimming pool. Guests are accommodated in two wood-and-stone houses. One has three apartments: *El Pajar* and *La Fresquera* (each sleeping 4) and the tropical *El Solarium* (sleeping 2), each with a kitchen and sitting room with an open fireplace. The house can also be rented out as a whole. There is also *El Cabrero*, a converted shepherd's hut with a bathroom, sit-in kitchen and open fireplace.

How to get there

By car: 5 km S of Plasencia, take EX108 W towards Coria and Moraleja. Just before Coria, take EX109 NW towards Moraleja and then Perales del Puerto. About 4 km N of Perales go west on EX205/C513 N towards Hoyos and Acebo. In Acebo, ask for directions (another 1.5 km). **Public transport:** Daily ENATCAR buses from Bilbao, Irún, San Sebastián, Pamplona, Vitoria, Burgos, Valladolid and Salamanca. From Madrid, AUTORES bus to Acebo. Then taxi or call the farm to be picked up.

81. El Molino. Parador de Ti Mismo

Luiz Lázaro & Eva Menéndez
Ctra Acebo - Ciudad Rodrigo km 1.5,
10857 Acebo, Cáceres
Tel: 927-19 30 39 / 927-51 44 79
Fax: 927-51 42 14
fundacion@ecotopia.es www.ecotopia.es
Open all year Language: E, GB
E pp B&B 15; full board 30; dinner 7; VAT 7%

Mill and surroundings

El Molino, an old water mill, is perched on the bank of the crystal clear Rivera de Acebo, a river that runs through a scenic valley. The mill is surrounded by 3 ha of orange groves, and 5 ha

of olives. The old buildings have been converted into a home for Luiz, Eva and their children. They are very environmentally aware, and brimming with ideas. Luiz is involved in several environment-friendly projects, among which are sustainable building and the creation of a biomass factory, intended to provide the entire region of Extremadura with clean energy.

Behind their house, Luiz and Eva have built simple, charming guest lodgings, using environmentally friendly construction materials, opening on to an enclosed courtyard. The furniture comes from Luiz and Eva's own carpentry workshop. The guests sleep on futons, and the lodgings use only solar and hydraulic energy. No smoking is permitted inside. Basic washing and toilet facilities are situated 10 m away (one room has its own facilities). There is a common sitting/dining room. Use of the kitchen is restricted to the low season. Breakfast is included in the charge. Luiz and Eva sell organic produce from their own vegetable garden; other local organic products are available in Acebo (2 km). Guests who wish to assist Luiz and Eva with farming, building or carpentry may do so by arrangement with their hosts. Other activities involving environmental awareness, health, recreation and personal development are available.

El Molino is situated in a landscape of rolling hills and indigenous chestnut and oak forests. Nestled between the rocks in the river you will find some excellent swimming spots. In winter, when the water is too cold for swimmers, the river is a source of drinking water. This area is ideal for walking and riding.

How to get there

By car: 5 km S of Plasencia, drive W towards Coria and Moralejas (EX108). Just S of Coria, drive N towards Moraleja and Perales del Puerto (EX109). About 4 km past Perales turn W

(EX205/C513) towards Hoyos and Acebo. In Acebo, ask for directions to El Molino (approx. 2 km). **Public transport:** ENATCAR buses depart daily from Bilbao, Irún, San Sebastián, Pamplona, Vitoria, Burgos, Valladolid and Salamanca. From Madrid, take an AUTORES bus. Taxis take you from the local bus stop in Acebo to El Molino (approx. 2 km).

82. Caserío de Fuente de Arcada

María Rosa de Torres Peralta
Ctra C513 km 23.5, 10893 Villamiel, Cáceres
Tel: 927-19 30 81 / 915-41 46 36
Fax: 915-47 03 22
Open all year Language: E, F, GB, I
€ pn B&B: 2p 69; Self-catering: 114

Farm and surroundings
This spectacular old farmhouse, set in a rustic and remote area, has a great view of rolling hills covered by chestnut and oak forest. Cows and horses graze the fields around the house. The outbuildings, chapel, springs, beautiful orchard and labourers' cottages around the main house give the property the air of a small village. There is a vegetable plot, and an olive orchard for the production of organic olive oil (also for sale). Maria also makes jam and wine.
  The luxurious guest rooms were lovingly decorated by the hostess herself. There is a spacious sitting room and a stylish dining room. Maria prepares breakfast (included) and also cooks dinners for large groups. The peace and quiet is interrupted now and then by an electricity generator. Some 100 m from the main house you will find a new, stylish vacation home for up to 6 guests, with a fireplace and a terrace: this is Casa del Caminito. The farm has

bicycles for hire and there is riding nearby.
  The region abounds with good walking trails (guided excursions are also available), beautiful rivers that are great for swimming, and various mediaeval villages.

How to get there
By car: 5 km S of Plasencia, drive W on the EX108 towards Coria and Moraleja. Just S of Coria take the EX109 N towards Moraleja and Perales del Puerto. Some 4 km past Perales, take the EX205/C513 W towards Hoyos and Valverde de Fresno. At km marker 23.5, turn right at the sign to Caserío de Fuente de Arcada. Drive another 200 m down this unpaved road. **Public transport:** ENATCAR buses depart daily from Bilbao, Irún, San Sebastián, Pamplona, Vitoria, Burgos, Valladolid and Salamanca. From Madrid, take an AUTORES bus. In Villamiel*, take a taxi or call to be picked up.

83. La Huerta de Valdomingo

Carlos Donoso
San Martín de Trevejo, Cáceres
Tel: 927-51 31 30 / 927-14 17 24
Mob: 689-40 07 50 ecogat@inicia.es
Open all year Language: E, P, F
Self-catering pn: €108(ls)–€144(hs); VAT 7%

Farm and surroundings
This small farmhouse sits on top of a hill at the edge of the village of San Martín de Trevejo. It has an excellent view of the valley, and there are wonderful red sunsets. In the opposite direction, guests look out over the rooftops of the picturesque village. The farm is surrounded by its own orchards and vegetable garden. Carlos and Carmen's older relatives work the land

in the traditional (organic) way. All year round this subsistence farm produces fruits, vegetables and nuts: oranges, potatoes, tomatoes, beans, almonds and more. Wine and olive oil are also made from the farm's own produce. The house is made of natural stone and solid wood. Its interior is rustic. There are four guest rooms, a bathroom, a toilet, a sitting room with an open fireplace, a fully equipped kitchen and a charming terrace. A clear mountain stream with a natural swimming pool is just 100 m from the house. Carlos and Carmen run an organic store in Acebo, which also offers various classes. Nearby you will find chestnut and oak forests, hills and Mount Jálama.

How to get there

By car: 5 km S of Plasencia, take the EX108 W towards Coria and Moraleja. Right before Coria, turn N on to the EX109 towards Moraleja and Perales del Puerto. About 4 km past Perales, take the EX205/C513 W towards Hoyos and Valverde de Fresno. Approx. 8 km past Hoyos, turn right towards Villamiel and San Martín de Trevejo. Check in at Casa Zoila on the Plaza Mayor. **Public transport:** ENATCAR buses depart daily from Bilbao, Irún, San Sebastián, Pamplona, Vitoria, Burgos, Valladolid and Salamanca. From Madrid, take an AUTORES bus.

84. Finca El Cabezo

Miguel Muriel García
10892 San Martín de Trevejo,
Cáceres
Tel: 927-19 31 06 Mob: 689-40 56 28
correo@elcabezo.com
www.elcabezo.com
Open all year Language: E, F, GB
€ *pn B&B: 2p 69-83; extra bed 22; B!*

Farm and surroundings

This 19th century farmhouse stands in a hilly, lush landscape alongside a quiet road to Portugal. El Cabezo is in transition to organic production of olive oil, and the farm also has 90 cows. Guests are most welcome to assist the owners and their two daughters, who follow environmentally friendly principles.

This typical Extremadura hacienda has large, comfortable rooms. Breakfast is included in the room price. Guests may prepare hot meals in the common kitchen. There is a patio, a large sitting/dining room and a delightful courtyard with an oven and a lookout tower. All guest rooms are stylishly decorated and include a bathroom with toilet. Three of the rooms have an open fireplace, and one has a sitting room. Guests can hire bicycles and receive free tourist information on the premises. Children can play safely here.

The farm is situated in the western part of the Sierra de Gata, a mountain range with pine trees, oaks and natural swimming holes. In the vicinity you will find pretty mountain villages built in traditional local style. The villages each have their own trade; basket weaving in San Martín de Trevejo, woodcarving in Valverde de Fresno and lace-making in Acebo.

How to get there

By car: 5 km S of Plasencia, take the EX108 W towards Coria and Moraleja. Right before Coria, turn N on to the EX109 towards Moraleja and Perales del Puerto. 4 km past Perales, take the EX205/C513 W towards Hoyos and Valverde de Fresno. At km marker 28.8, turn left at a wooden sign reading Casa Rural El Cabezo. **Public transport:** ENATCAR buses depart daily from Bilbao, Irún, San Sebastián, Pamplona, Vitoria, Burgos, Valladolid and Salamanca. From Madrid, take an AUTORES bus.

Andalucia and the Canary Islands

Andalucia

Deer in the province of Córdoba

The landscape of Andalucia, Spain's southern-most region, is dominated by olive orchards and holm-oaks. But not all of Andalucia is rough and dry terrain: the region also has verdant nature parks and mountain ranges, such as the Sierra Nevada, south of Granada, which has peaks as high as those of the Pyrenees. At the southern foot of these mountains, you will find Las Alpujarras, which take their name from the Arabic *Al Busherat* (grassland). The Moors used this terraced land to grow their subtropical fruit. The eastern half of the Alpujarras are relatively dry. At Ugíjar, the terrain is still rough and yellowish. Further west, it gradually turns green with poplar, chestnut and walnut trees. Today, water still flows through irrigation channels once built by the Moors. There are many beautiful mountain villages in the area, which have preserved their characteristic folk building style. A prime example of this is Trevélez, Spain's highest village, with many shady plane and poplar trees.

Andalucia is also famous for its Moorish architecture, which reached its pinnacle in the stunningly beautiful Alhambra palace in Granada. Córdoba and Sevilla also have their share of Moorish monuments. Andalucia was occupied by the Moors longer than any other Spanish region: they arrived in 711 and did not leave until 1492, when they were ousted from their last stronghold in Granada. Numerous castles and citadels (*alcazares* and *alcazabas*) are a tangible reminder of this period of colonization. Smaller towns like Úbeda and Baeza (near Jaén), which are close to several ECEAT places to stay, also have interesting architecture. Both towns have beautiful plazas surrounded by Renaissance buildings.

Arrecife de las Sirenas, Cabo de Gata Almería

Parks

Andalucia's nature parks are less well known. North of Granada, between Cazorla, Beas de Segura, Villarodigo and Poco Alcón, you will find the Parque Natural de las Sierras de Cazorla, Segura y las Villas. This park encompasses many diverse ecosystems. Measuring over 214,000 ha of terrain, this is the largest protected area in all of Spain. Many mountain peaks are over 2,000 m high. This is also the source of the river Guadalquivir, which begins as a narrow

81

mountain stream, but as it makes its way to Sevilla, swells into the widest river in Andalucia. The park's vegetation and wildlife are incredibly varied. One unique plant of the region is the Cazorla violet. The mixed oak, holm and pine forests are home to deer, wild boar, squirrels and many other animals. Among the rarer species are the lammergeyer and the Spanish ibex. In more humid areas, you can come across otters, frogs, salamanders and water snakes. You may even see a golden oriole or a kingfisher. Parts of the park have restricted access and can only be visited with a guide. The park's Torre de Vinagre information centre, which includes a zoo and botanical gardens, are 17 km north of Cazorla, on the road to the El Tranco reservoir.

There are some parks in the area around Córdoba and also near Ronda, where you find the Sierra de Grazalema (50,000 ha) and Los Alcornocales. The Sierra Grazalema varies in elevation from 50 - 1,650 m. It is the western end of the Penibética mountain range, and has its own microclimate, with the highest rainfall in Spain. The area is full of caves, and is also home to the rare *pinsapo*, a silver fir that dates back to the Tertiary Era. The area's more common trees include cork-oaks and holly-oaks. The mountains provide a habitat for the mountain goat, roe deer, fox and numerous birds of prey including a large colony of vultures. You can explore the area by car (route descriptions are available) or walk one of several marked paths.

Spain's most famous (and infamous) national park is the Parque Nacional de Doñana at the Guadalquivir delta in the provinces of Huelva and Sevilla. Wildlife and environment there have been under threat since 1998, when a nearby Swedish-Canadian mining company badly contaminated the water and the soil. Since the disaster, everything possible has been done to clean the area up. It is uncertain whether the Doñana will ever fully recover, but there is hope. The area's salt marshes, dunes and pine forests provide shelter for the wintering flamingos, cranes and other aquatic birds. The area is also home to the Iberian lynx and the Egyptian mongoose. Paid access by guided tour only. Organized tours start from the visitors' centres in El Rocina and El Acebuche.

Canary Islands

The Canary Islands consist of seven large and several small volcanic islands, approximately 1,000 km south of the Iberian peninsula and some 130 km west of Morocco. The archipelago is unique in many ways. Although all the islands share cool trade winds and a mild, sunny climate - their perpetual spring has been famous since antiquity - each has its own distinct landscape. In a nutshell, the islands' history of human habitation seems to have begun with the Guanches. This mysterious, materially primitive but morally superior people tried to withstand the Spanish conquerors, but eventually lost the battle. From the 14th century, the Spaniards settled on the islands and started cultivating the land. They grew sugar cane, wine grapes and produced cochineal. Nowadays the main crops are bananas and tomatoes. Besides agriculture, nature is an important source of income. Walkers consider the islands a Mecca: a maze of footpaths leads through four national parks and over 100 other nature reserves and parks.

N° 101. Las Castañetas

Gran Canaria

The great differences in elevation on Gran Canaria account for the island's different microclimates. The southern half of the island, which is shielded by the mountains, is warmer than the subtropical northern half. Gran Canaria has a reputation for being a tourist trap. While it is true that the island

*Griffon vulture,
Sierra de
Grazalema, Cádiz*

attracts many people who come for its sunny beaches and pleasant nightlife in the south, those who venture further inland will embark on a voyage of discovery. The interior abounds with spectacular, unspoilt landscapes, featuring deep gorges, water reservoirs bordered by pretty vegetation, unparalleled vistas, craggy cliffs and dark green forests. The island offers an incredible variety of walks through unspoilt villages, green meadows, dense pine and deciduous woods in the north, and, once past the La Cumbre massif, through desert-like and subtropical scenery.

Many of Gran Canaria's plants exist nowhere else in the world. The island also has an abundance of flowering plants such as the hibiscus and bougainvillea. There are even a few indigenous animal species. Walking is highly recommended. The old paths connecting the villages have been restored.

The various towns and villages on the island are also worth visiting. There are clear differences from area to area. The north - the most fertile, and earliest cultivated and inhabited territory - has traditional villages and the historical town of Teror. The northern interior is characterized by the presence of numerous fruit trees and splendid gardens. And the south - originally the driest and quietest region - is now the busiest part of the island, with many tourist-oriented activities, ports and small fishing communities. The town of Telde is the south's most important historical landmark. Day trips to any of these areas are recommended.

Fuerteventura

Fuerteventura has miles of white, sandy beaches and crystal-clear turquoise waters, and far fewer tourists than you find on Gran Canaria. The island is great for swimming, sailing and windsurfing, and it has excellent diving and snorkelling spots. As there are only small differences in elevation on Fuerteventura, there is very little rain on the island, so in no way does it resemble green Gran Canaria: it has its own distinct landscape, with large expanses of golden dunes and many volcanic formations. Once densely wooded, it was deforested by the colonists. Imported goats ate what few trees were left, but the island's goat cheese is world famous. The resulting aridity sparked off a flourishing fishing industry. There are various fishing villages where you can eat great seafood, for example in El Cotillo and Corralejo.

As you make your way across the island, you will enjoy stunning desert-like landscapes with an occasional windmill. Make sure to stop in Betancuria, a village that looks as if time has stood still since the 15th century. Stay for a spectacular sunset and enjoy the sense of being one with nature. "Fuerteventura," Miguel de Unamuno once wrote, "is a spiritual rock."

85. El Cordonero & El Cedro

Urbana Fernández González
Maestra Adame 16, 21292 Fuenteheridos,
Huelva. Tel: 959-12 50 19
Open all year Language: E
*€ pn Self-catering: 45.18; 270 pw; extra bed
9.04; add. cleaning*

Cottages and surroundings
These 2 traditional cottages are situated in the
Sierra de Aracena y Picos de Aroche, a regional
park in NW Andalucía and one of the greenest
areas in the entire region. There are many holly-oaks and cork-oaks there, and large tracts of
land are being reforested with pine and chestnut. The area is famous for its organically bred
Iberian Patanegra pigs, and for delicious ham
dried in the pure mountain air. The proprietress makes decorations from dried flowers
and sells them at an affordable price.
    There are two lodgings for hire. *El Cor-
donero*, a traditional farmhouse, is a simple
one-storey cottage with no electricity. It has a
living room with a fireplace, kitchen, bathroom
and three double bedrooms. Outside there is a
large terrace where you can enjoy the shade of
a large chestnut tree. *El Cedro* is an apartment
in an old building with a spacious terrace under a tall cedar. Inside you will find a living
room with a fireplace, kitchen and three bedrooms all furnished with antiques, and a modern bathroom. The two lodgings share a tank
that stores clean spring water.
    The park is renowned for its riding, walking
and cycling trails. Aracena is a beautiful mountain village with the ruins of a castle and
church that belonged to the Knights Templar.
Nearby you will find the Gruta de las Maravil-

las, a natural cavern over 1 km deep, with large
chambers, lakes, stalactites and stalagmites.

How to get there
By car: The owner sends information after
guests book their stay. She escorts you person-
ally to the lodging. **Public transport:** Train from
Huelva or Mérida to Cortejana/Canaleja, then
the bus to Fuenteheridos.

86. Las Navezuelas

Luca Cicorella
Apartado 14, 41370 Cazalla de la
Sierra, Sevilla. Tel: 954-88 47 64
Mob: 630-11 55 70 Fax: 954-88 45 94
navezuela@arrakis.es
www.arrakis.es/~navezuela
*Open all year Language: E, GB, F, I
€ pn B&B: 2p 57-69(ls) 60-75(hs);
Self-catering: 81-90(ls) 87-96(hs); extra bed
15(ls)-18(hs); lunch & dinner 13.50; VAT 7%*

Farm and surroundings
Situated in a valley of the Sierra Morena, this
36 ha farm is surrounded by green pastures
bordered by holly-oaks, cork-oaks and low
olive trees. Organic fruit is grown on 1 ha of
the land. The historical village of Cazalla de la
Sierra is 3 km away.
    The 16th century olive mill has been convert-
ed to house the guest lodgings. Guests can
choose from single and double bedrooms with
their own bathroom, studio-type rooms, or a
detached apartment. There is an opportunity
to meet and mingle with other guests in the
common sitting rooms. In the big kitchen local
specialties are prepared using mainly home-
grown products. Guests can participate in pot-

tery workshops or learn how to make fruit preserves. There are several wildlife observation posts.

Cazalla de la Sierra has a visitors' centre with information about the Sierra Morena nature reserve.

How to get there

By car: From the village of El Pedroso on the A432, follow the signs to Cazalla de la Sierra. Approx. 3 km before Cazalla (at km marker 43.5), turn right at the sign to Las Navezuelas. The farm is 400 m down this unpaved road. **Public transport:** Train from Mérida or Sevilla to Cazalla de la Sierra (train station is outside village). Bus to Cazalla de la Sierra, then taxi.

87. El Berrocal

Carmen Centeno Guerra
Ctra. Real de La Jara km 1, 41370 Cazalla de la Sierra, Sevilla
Tel: 954-88 44 22 Mob: 650-69 91 86
Open all year Language: E
B&B & self-catering pppn: €27;
lunch & dinner €12.02-€15.02; VAT 7%

Farm and surroundings

This traditional *hacienda* is near the village of Cazalla de la Sierra, in the middle of the Sierra Norte nature reserve (part of the larger Sierra Morena). It is a family farm, with cattle and horses. Part of the main building is used by the proprietress, the rest is for guest lodgings (5 double rooms and 2 bathrooms). Some of the farm's other buildings have been converted into 8 self-catering apartments for 2-4 guests, each with its own particular characteristics. All have spacious bedrooms, a living room, kitchen

and bathroom. There is a swimming pool on the premises. The owner serves meals made with home-grown products. Guests are welcome to help do the daily chores on the farm.

Horse-back riding (day trips and classes), pottery, cheese and jam-making courses are offered at the farm, and you can swim near a waterfall 20 km from the farm.

How to get there

By car: From Sevilla, take the A431 N. Just west of Cantillana, take the A432 to El Pedroso and Cazalla de la Sierra. Just before the village turn left towards El Real de la Jara untill you see the sign to the right to El Berrocal (SE179). **Public transport:** Train from Mérida or Sevilla to Cazalla de la Sierra (the train station is outside the village). Take a bus to Cazalla de la Sierra, then a taxi to the farm.

88. Cañada de los Pájaros

Maribel Adrián Dossío
Ctra Puebla-Villafranco km 8, 41130 La Puebla del Río, Sevilla
Tel/fax: 955-77 21 84 Tel: 950-26 50 18
lospajaros@donana.es
Open all year Language: E, F
Self-catering pn: from €42; extra bed €9

House and surroundings

A great holiday spot for nature and bird lovers. The owners have spent 10 years cleaning up this 6 ha area: they created large ponds and transformed this former gravel pit into a bird sanctuary. The proprietors' house is in the middle of the park. Under the same roof is the self-catering apartment, with a living room and kitchenette on the ground floor. One storey up is the double bedroom and a bathroom. The owners, who are also rangers in the sanctuary, are happy to show you the surrounding area and describe the ecosystem.

The house is ideal for peace and quiet. It is also an excellent base from which to explore SW Andalucía. Sevilla is just 30 minutes away by car. It is a 45-minute drive to the town of Matalascañas on the Atlantic coast and 1.5 hours to Portugal. Doñana national park is just 15 km away.

How to get there

By car: SE660 from Sevilla to Coria del Río. Pass Coria and La Puebla del Río. Then SE659 (Villafranco/Aznalcázar) following signs to Reserva Natural Cañadas de Los Pájaros. After about 350 m of unpaved road you reach house. **Public transport:** Train or bus to Sevilla, then taxi or another bus to La Puebla del Río, and taxi.

89. Finca Paquita

Francisca Benítez Barranco
Bocaleones s/n, 11688 Zahara de la Sierra, Cádiz. Tel: 956-23 40 42 Mob: 629-98 42 55
Open all year Language: E, D
€ *pw Self-catering: 2p 300.50; 4p 480.80; VAT 7%*

Cottages and surroundings

Finca Paquita is a traditional Spanish residence situated in the northern part of the Sierra de Grazalema nature reserve. The house is surrounded by a large shady garden, terraces at different levels, fruit trees and a vegetable plot. Doña Paquita welcomes you to her carefully tended home.

The 4 guest cottages around the main build-

ing sleep 2-4 people each. The grapevine-covered patios are sheltered from the wind and equipped with barbecues. Tucked away between swaying cypress trees and exuberant Mediterranean flowers is a swimming pool. A grassy area surrounding the pool has reclining chairs for sunbathing. From here you have an incredible view of nearby Zahara de la Sierra. The *finca* has plenty of tourist information.

Local attractions include the spectacular Sierra de Grazalema nature reserve. The reserve has several information offices and good road signs. However it has very strict access rules to protect the many rare birds there (ospreys, vultures, etc.). To walk in the Gargante Verde (10 km away), you must make reservations. Heavy annual rainfall and expansive underground water reservoirs give Sierra de Grazalema a lushness uncharacteristic of Andalucía. The reserve has excellent walking, cycling and riding paths. Conservationist groups offer guided excursions (1 day, 2 days or longer). The charming village of Zahara, perched high above the area, has authentic-looking tapas bars and the remains of a Moorish castle. Zahara, Benamahoma and Grazalema are among Andalucía's prettiest villages, on the Ruta de los Pueblos Blancos.

How to get there

By car: From Algodonales on the A382, go S towards Zahara. After reservoir, turn right on to CA5232 towards Prado del Rey. After about 1 km, cross bridge, then road bends to the right. Immediately after bend, there is sign for Finca Paquita: turn right on to red dirt road and drive another 0.3 km. (NB: there is another red dirt road just before bridge - ignore this). **Public transport:** Train to Ronda, then bus to Zahara (daily at 2 p.m.). Or take bus from one of the large cities to Algodonales, then taxi.

90. Camping-Cortijo La Jaima

José A. Amaya
Ctra Prado-Arcos de la Frontera km 1.3,
`11660 Prado del Rey, Cádiz
Tel: 956-72 32 35 julia@viautil.com
Open Mar 21st-Oct 21st Language: E, GB
€ *pn Camping: tent & caravan 2.40; adult 2.55; child 1.40*

Farm and surroundings

Lambs, chickens and turkeys are raised on this traditional farm, from where there are wonderful views of the surroundings. The farmers have two children.

There is a spacious campsite, where the pitches have some privacy and trees provide shade. Campers can use the common sitting room and swimming pool. There is also a self-catering cottage big enough for 10 guests. Bicycles are for hire. Groups of 6 or more can take a course in ceramics.

Cortijo La Jaima is great if you like walking, riding and cycling. It is on the Ruta de los Pueblos Blancos, a route dotted with white stuccoed villages with narrow alleys and balconies full of red geraniums. El Bosque (7 km from the farm) has a visitors' centre which provides information about the regional park. In late July there are flamenco festivals.

How to get there

By car: From Sevilla or Málaga, drive to the A382 (Antequera-Jerez). In Villamartin, take the A373 to Prado del Rey. Drive through the village and take the CA5231. At sign indicating 1.3 km, follow signs to campsite. **Public transport:** Regular bus service from Sevilla, Jerez and Cádiz. Take bus to Ubrique and get out in Prado del Rey. Walk the remaining 2 km or take a taxi.

91. Casa Montecote

Gisela Merz & Rainer Wiessmann
Lugar de La Muela 200, 11150 Vejer de la Frontera, Cádiz. Tel/fax: 956-44 84 89
montecote@terra.es
www.casamontecote.de.vu
Open all year Language: E, GB, D, F, I
€ pn Self-catering: 38-60(ls) 51-85(hs);
240-368(ls) 350-520(hs) pw; massage 15

Farm and surroundings

Casa Montecote is a small organic farm at the edge of a hamlet called La Muela. It has a wonderful view of the fields and mountains around Vejer de la Frontera and the Janda lagoon. You can relax in the swimming pool or sauna on this green oasis, or help out on the farm.

The guest house and various apartments each have a beautiful private terrace. One of the apartments is wheelchair-accessible. You can buy the farm's own vegetables, milk, eggs, bread and other products. Children will enjoy the small gym and many outside games.

There are plenty of activities nearby. The beach is 8 km away. Visit the ruins of Bolonia and the cities of Tarifa, Cádiz and Jerez. From Arcos de la Frontera, you can walk the Ruta de los Pueblos Blancos.

How to get there

By car: On the N340 from Cádiz to Algeciras, exit at km marker 27.5 and head towards La Muela on the CA9008. In La Muela, look for the Venta Rufino bar. There, take the unpaved road uphill. After 700 m you will see a sign for Casa Montecote farm. **Public transport:** Train to Cádiz, then the bus to Vejer-Barbate. Get out at La Muela.

92. Casas Karen & Fuente del Madroño

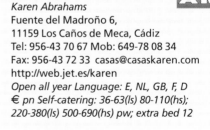

Karen Abrahams
Fuente del Madroño 6,
11159 Los Caños de Meca, Cádiz
Tel: 956-43 70 67 Mob: 649-78 08 34
Fax: 956-43 72 33 casas@casaskaren.com
http://web.jet.es/karen
Open all year Language: E, NL, GB, F, D
€ pn Self-catering: 36-63(ls) 80-110(hs);
220-380(ls) 500-690(hs) pw; extra bed 12

Houses and surroundings

Los Caños de Meca is a fairly unspoilt location with six white sandy beaches, cliffs, springs, caves and an immense blue sea. It is near the Straits of Gibraltar, with a view of the lighthouse at Cape Trafalgar and the Moroccan mountains across the sea. The surrounding hillsides are covered with umbrella pines. Casas Karen, five separate vacation homes, are just five minutes' walk from the beach.

The *casas* are built in traditional Andalusian style with Moroccan influences. Each has its own open fireplace or wood stove, and a spacious terrace with a Mexican hammock. Surrounded by broom and mimosa, some have a view of the hills and evergreen woods, others look out over the sea. You can also spend the night in the romantic *choza*, a straw hut typical of the region. Organic vegetables can be ordered from a nearby farm. Reiki, yoga and aromatic massages are available. There is plenty of information about the surroundings.

Casas Karen borders on the Acantilado y Pinar de Barbate nature reserve, where there is a great variety of plants and wildlife. It is a bird watchers' paradise, and excellent for riding. Nearby bars and restaurants serve wonderful fish dishes. The sea is great for sailing, diving, windsurfing and swimming.

How to get there

By car: On N340 (Cádiz-Tarifa), exit towards Vejer at km marker 35. At roundabout immediately turn right to Los Caños de Meca. After 10 km (500 m after exit to Faro Trafalgar), turn left on to dirt road with sign reading *apartamentos y bungalows Trafalgar*. After 400 m turn right towards Fuente del Madroño. After 80 m, turn right into wooden entrance gate. Reception in 2nd straw hut (if you arrive outside business hours call mobile number). **Public transport:** Buses from Cádiz and Barbate pass through Caños de Meca (2x daily). Or bus to La Barca de Vejer, then taxi (ask for Ramón: he knows where Casas Karen is).

93. Casarosa

Rias Kanon
Lista de Correos, 29390 Estación de
Gaucín-Colmenar, Málaga
Tel: 952-11 70 32
Open all year Language: E, NL, GB
Lodging pp: full board €21; B!

Farmhouse and surroundings

This converted farm is right in the middle of a very old cork-oak forest. The property is dotted with fruit trees including peach, apricot, orange and almond. A natural spring supplies the farm with water. Solar panels are used to heat the water and to generate electricity. There are 2 bedrooms and a wooden cabin for rent. Rias, your host, prepares vegetarian meals with home-grown organic produce and bakes fresh whole grain bread every day.

North of Casarosa is Sierra Grazalema nature park, where Europe's largest bat colony spends the winter. To the south, you will find Los Alcornocales-Sierra del Aljibe, an unspoilt nature reserve rich in wildlife. Among the animals to be found here are the civet, mountain cat, deer, wild boar, otter, European turtle, swift and various birds of prey. The beach is a 40 km drive through mountains. Along the way you will see several white stuccoed villages that reveal their Moorish roots, and where you will always find an old castle or fort.

How to get there

By car: From Ronda, take the A369 S until approx. 2 km past the road to Gaucín and Manilva. Turn right on to the MA512 to El Colmenar. Further details available when you book your stay. **Public transport:** Train from Granada or Córdoba in the direction of Algeciras. Get out at Estación de Gaucín-Colmenar. From the station, walk the remaining 1.5 km to the farm.

---

94. Finca La Mohea

Ruth Bond & Rory Corcorán
Finca La Mohea s/n
29492 Genalguacil, Málaga
Tel: 952-11 71 21
lamohea@costasol.es www.lamohea.com
Open Mar 21st-Nov Language: E, GB
B&B pp: €12; half board €18;
lunch & dinner €6

Farm and surroundings

Finca La Mohea (elev. 500 m) is surrounded by cork-oaks, holm-oaks and chestnut trees. The owners make a living from organic farming and permaculture, selling whatever produce they do not need for their own use. They also keep goats, chickens and geese. The family moved here some 12 years ago, and their respect for nature has greatly increased the local biodiversity. This is an excellent holiday spot for guests who love the countryside and who are curious about permaculture and organic farming. The house has 2 double bedrooms for guests and basic facilities: an outdoor bathroom and a compost toilet. Solar panels provide electricity for this no-frills farmhouse. Your hosts serve tasty, healthy meals.

The surrounding valleys and wooded mountains are dotted with small family farms with terraced orange groves and other orchards. Part of the forest is cultivated for sweet chestnuts and cork, while the higher mountain forests are wilder. The highest peak is Los Reales, at 1,450 m. Indigenous plants and wildlife abound here. Eagles soar over the

trees, which provide cover for wild boar, deer and mountain goats.

How to get there

By car: Genalguacil is situated inland from Estapona (halfway between Marbella and Gibraltar). In Genalguacil, take the high road through the village. This turns into an unpaved road towards Puerto de Peñas Blancas. Drive 6 km to a river, cross the river and continue for 1 km. At the next crossroads turn left. The farm is 1 km down this road, on the left-hand side.
Public transport: Take a bus from Ronda to Genalguacil (every afternoon). Call in advance for pickup service.

---

95. Finca Fantástico

Fernando Picasso
Tenda la Gavia, 29510 Álora, Málaga
Mob: 639-28 98 94
fantastico_gallery@yahoo.com
Open all year Language: E, GB
€ pn Camping: adult 9.02; child 5.95;
camper van 3.01; Self-catering: 36.06-54.01;
252.43-378.64 pw; extra bed 18.03;
breakfast 6.01; dinner 15.03

Farm and surroundings

Finca Fantastico (elev. 700 m) is an organic almond farm situated on 10 ha of unspoilt Spanish countryside, deep in the Penibética mountains. Peace and quiet are plentiful here, in the rolling hills covered with olive and carob trees. The farmhouse is over 500 years old and has been lovingly restored over the past decade. It retains many original features such as beamed ceilings and an old fireplace. The campsite has 47 pitches: 37 for tents and 10 for small camper vans. There is also a cottage (max. 5 guests), which can be divided into 2 apartments and there is a group lodging large enough for 10 people. Breakfast is optional and is served outside, in a large, shady barbecue/bar area. The farm owner is a painter and photographer, who gives workshops in painting and sculpture. Guests are welcome to help out on the farm.

The surrounding area is great for walking, cycling and bird watching. The village of Álora bears clear traces of the Roman, Visigoth and

Moorish civilizations. The nearest swimming spot is 15 km away.

How to get there

By car: From Málaga, take the half-hour drive down the new A357 to Estación de Cártama. Ask in the village for 'La Cooperativa de Almendra' (the biggest organic almond co-operative in southern Spain). From there, you will find signs to Finca Fantastico (another 12 km). If you have trouble finding the farm, do not hesitate to call for directions. **Public transport:** From Málaga take a train or bus to Estación de Cártama or Álora. Call for pick-up service (€ 9.02).

---

96. Casa de Elrond

Mike & Una Cooper
Barrio Seco s/n
29230 Villanueva de la Concepción, Málaga
Tel/fax: 952-75 40 91 elrond@mercuryin.es
Open all year Language: GB, E
B&B pn: 2p €40; dinner €15

Cottage and surroundings

When Mike and Una Cooper renovated Casa de Elrond they took care to maintain its traditional Spanish style (whilst adding a few British touches). Their bed and breakfast has 3 rooms. The doubles are big enough for a family of 4, and every room has its own bathroom. In win-

---

ter, you can read a good book by the living room fireplace or socialize with other guests. A shady terrace with a coastal view, a place to sunbathe and a swimming pool await you outside. The pretty front garden has various fruit trees (lemon, persimmon, grapefruit etc.) and the orchard has olive and almond trees. Dinners (vegetarian only) are prepared on request.

The surroundings are excellent for walks, offering superb panoramas and a wealth of flora and fauna. The nearby El Torcal nature park (visitors' centre, marked footpath) is well worth a visit. Villanueva is a characteristic Andalusian village where time seems to have stood still. Antequera is 14 km away, while Málaga and the Mediterranean are 35 km to the south.

How to get there

By car: From Málaga take the N331 in the direction of Antequera. Get off motorway at exit Casabermeja and continue on the MA436. After about 8 km, turn right towards Villanueva de la Concepción. Once there, turn left at a T-junction at the top of a hill (to your right will be the town hall, 'Ayuntamiento'). Casa de Elrond is 3 km down the road on your right. **Public transport:** Train from Granada or Sevilla to Antequera. From there, take a taxi or call in advance to be picked up.

---

97. Las Cañadas

Juan Antonio García Olmedo
Portugalejo s/n, 29532 Mollina, Málaga
Tel: 952-74 13 28 Mob: 610-71 61 72
Fax: 952-74 11 67
Open all year Language: E, GB
Lodging pn: 2p €34.26; VAT 7%

Farm and surroundings

Las Cañadas is a horse-breeding farm just out-side Mollina, overlooking the Antequera low-land. In the distance you can see the Torcal mountains. Bed and breakfast is available, but you can also hire the entire house. There are 4 double bedrooms, each with its own bath-room, underfloor heating and solar heated wa-ter. There is also a large living room with a fire-place, and a kitchen. You can seek relief from the midday heat in the swimming pool or on the spacious, shady patio.

The horse stables are across from the house. The owner arranges riding, walking, cycling, caving and rafting tours. Your help is welcome in the vegetable garden. The house is centrally located; it is a 1.5 hour drive to Sevilla, Málaga, Córdoba and several other historic sites. Closer by (a 20 min. drive from the farm) you will find the ancient town of Antequera, El Torcal na-ture reserve (gorges, natural bridges, caves and many birds of prey) and the Laguna de Fuente Piedra bird sanctuary with its flamingo colonies.

How to get there

By car: From the Sevilla-Granada motorway (A92), take one of the exits to Mollina. From there, call to be escorted. The farm is very hard to find. **Public transport:** Train from Granada or Sevilla to Antequera, then a taxi to Mollina. Arrange in advance to be picked up.

98. Finca Cuevas del Pino

Pilar del Pino López &
Ignacio Amián
14710 Villarrubia, Córdoba
Tel: 957-32 70 40 Tel/fax: 957-45 83 72
alhena@teleline.es cuevasdelpino.es.fm
Open all year Language: E, GB
€ pw Self-catering: 330-420(ls) 451-540(hs)

Farm and surroundings

Cuevas del Pino is a typical Andalusian farm with a splendid view of the Sierra Morena and the fertile Guadalquivir valley. This is a peace-ful, wooded spot. The 16 ha of farmland is partly organic; the rest is in transition to organ-ic agriculture. There is an old Moorish quarry on the property.

There are 2 self-catering lodgings: a cottage and an apartment sleeping 4 and 7. Both are nicely decorated and have their own living room, kitchen, bathroom and large garden. The cottage has a fireplace. The garden is big enough to pitch 2 tents, but there are no spe-cial facilities for campers. Call ahead if you want to camp out. The farm has a swimming pool, and horses which you may ride. There is also a studio where you can take courses in wa-tercolour painting and etching. Your help is welcome on the farm. There is plenty of infor-mation about long and short walks. Just 14 km away you will find Córdoba, with its famous Moorish mosque, one of the few synagogues that survived the Moorish invasion, and many colourfully tiled patios. The Almodóvar del Río castle is just 7 km away.

How to get there

By car: From Córdoba, A431 towards Palma del Río. Villarrubia is at km marker 12. Pass Citroën dealership and turn right before petrol station. Take this narrow road for 2 km, until the end. Turn left and drive another 100 m to entrance to Cuevas del Pino. **Public transport:** Train to Córdoba, then bus to Villarrubia. From there, walk (2 km) or take a taxi.

99. Villa Matilde

Merche Coco Pérez & Roland Wassenaar
Viñas de Peñallana 302; Ctra Pantano
Jándula km 3, 23740 Andújar, Jaén
Tel: 953-54 91 27 (020-693 73 15 in NL)
villamatilde@amsystem.es
www.infoandujar.com/villamatilde
Open all year Language: E, NL, GB, F, D
€ *pn Lodging: 1p 20.50-26.50;*
2p 26.50-32.50; 3p 33.50-39.50;
breakfast 3.60; lunch 7.20; dinner 9.60; B!

Manor house and surroundings

Villa Matilde provides a refuge for nature lovers seeking peace and quiet. The stylish manor house is surrounded by 3 ha of wooded property. The building dates back to 1926 and still has part of its original interior. Guests may use the common sitting/dining room (with a fireplace), take a dip in the swimming pool, sunbathe on the adjoining deck, or enjoy the shade of large pines and oaks in the garden. On request, your host prepares (vegetarian) meals using home-grown organic produce. Besides accommodating guests (there are 7 rooms sleeping 1, 2, and 3 persons), the owners run an ecology education centre for Spanish children, and offer various courses (Spanish language, painting, yoga). For those who love old legends there is plenty to discover: for instance the annual procession to mark the apparition

of the Virgen de la Cabeza in 1227. Dogs are allowed if they stay outside. Bikes can be hired.

Nearby you will find the Parque Natural Sierra de Andújar, a splendid park in the Sierra Morena. There are also two large water reservoirs with beaches. Many herbs grow wild in the park, such as myrrh and rosemary. The wildlife includes birds of prey, deer, wolves, lynxes and wild boar. The mountains have good trails for walking and mountain biking.

How to get there

By car: N-IV (E5) Madrid-Córdoba. Take exit 321 near Andújar. At 2nd roundabout turn right towards Viñas De Peñallana (14 km). After Los Pinos restaurant, turn right heading for Embalse de Jándula. After 3 km, turn right at the sign for Villa Matilde. **Public transport:** Direct service by bus or train from Madrid, Málaga, Sevilla, Córdoba or Jaén to Andújar. The bus from Granada travels via Jaén. From Andújar take a taxi or call to be picked up (at a fee).

100. Eco-turismo La Pendolera

José González Miras
La Pendolera s/n, 23380 Siles, Jaén
Tel: 953-12 60 35 / 953-12 61 15
Mob: 677-87 21 41
Open all year Language: E, GB
€ *pn Lodging: 1p 12; 2p 24; 3p 27; extra bed*
6; Self-catering: 60.01-90.15; 360.60-520 pw;
extra bed 60 pw; Groups pp: 9; breakfast &
lunch-pack 1.80; lunch & dinner 6.01

La Pendolera and surroundings

La Pendolera is a hamlet of just a handful of farmhouses in the northern part of Sierra de Segura, Las Villas and Cazorla nature park. The little farming community is situated amidst the woods and olive orchards. The owner and his friends have restored La Pendolera bit by bit, taking care to leave its original character intact. In the summer, groups of schoolchildren stay there; they cook together and bake bread in the brick oven. You are welcome to join them. Meals are made mainly with home-grown vegetables. José knows the nature reserve like the back of his hand. He works in Siles, at the Tienda Verde, which provides information about the park. José also organizes field trips for youngsters.

The hamlet is an ideal point from which to explore the surroundings by foot, for instance to find the source of the Río Mundo while enjoying the plants and wildlife along the way. Siles is a village with a rich Moorish history that is tangible in many of the buildings still standing today. Be sure to visit the Sierra de Génave co-op, which pioneered organic olive growing.

How to get there

By car: From Siles, take the A310 towards La Puerta de Segura. After 4 km, turn right and follow the signs to La Pendolera. The lodgings are about 1 km down the road, on the right side. **Public transport:** Bus to Siles. Arrange in advance to be picked up.

101. Las Castañetas

Cees van den Bogaard & Riecky Berends
Ctra Aguacebas km 24
23300 Villacarrillo, Jaén
Tel: 953-12 81 51 Mob: 686-39 25 29
Open all year; camping Jun 24th-Oct 10th
Language: NL, E, GB, D
€ pn Tents & caravans to rent 30;
Self-catering: 45.18

House and surroundings

Las Castañetas nestles among rocky mountains in the middle of the Sierra de las Villas. The peace and quiet in this idyllic place is broken only by birdcalls and wind blowing through the pines and holly-oaks. In 1991, Riecky and Cees moved here with their children, Vivian and Julien. Wild boar, deer, lynxes and wild cats roam in the mountains. In the summer, home-grown organic produce and basic foodstuffs

are sold in the small shop run by Vivian. There is a cottage for 5-8 people, caravans and fully furnished hire tents - each on its own beautiful pitch. You may not use your own tent or caravan. Be sure to bring enough provisions to last for your entire stay. Each accommodation has its own outdoor kitchen and lovely, cool spring water. The water from the pretty, chlorine-free swimming pool is also used to irrigate the fruit trees. The owners organize arts and crafts courses for groups of over 5 people.

The area is ideal for walking. Hostess Riecky can supply you with a description of 10 trails nearby, and Vivian can direct you to all the springs, waterfalls and look-outs in the vicinity.

How to get there

By car: From Villacarrillo, take the C323 to Mogón. From there, head towards Cueva del Peinero and then Sierra Las Villas nature reserve. In the park, follow signs for Pantano/Embalse de Aguascebas. 300 m past km marker 24 is a sign for Las Castañetas on the right side. Follow this narrow road for 3 km through the mountains. Driving from Cazorla, head for Santo Tomé on the JV7101. After 11 km, turn right on to the JV7102 towards Chilluevar. From there, turn left towards Pantano del Aguacebas and Sierra Las Villas. Continue as described above. **Public transport:** Take a Eurolines bus to Bailén or a train to Linares-Baeza. Arrange in advance to be picked up there (3,000 pesetas per person round trip; no charge for guests staying over 10 days).

102. Camping-Cortijo San Isicio

Jeanne & Jo Driessen
Camino San Isicio s/n, 23470 Cazorla, Jaén
Tel: 953-72 12 80
Open Mar-Nov 10th Language: NL, E, GB, D
€ pn Camping: tent 2.60-3.30; caravan 3.40;
camper van 3.95-5.15; adult 3; child 2.50;
Self-catering: 2p 40; 4p 45

Campsite and surroundings

Cortijo San Isicio is an organic apiary and fruit farm. Organic products such as honey and its by-products, candles and essential oils are for sale. The campsite is located in a quiet part of an area of great natural beauty, close to Sierra de Cazorla nature reserve and 2 km from the pretty mountain village of Cazorla. The campsite consists of terraced fruit orchards. You can pitch your tent in the lush grass in the shade of a tree. There is a barbecue area, a place to build campfires, and a lovely swimming pool. Just above the campsite is a cosy cabin for 4 guests, with a stunning view of Cazorla and the surrounding mountains. Hostess Jeanne provides route descriptions of several walks which take you directly from the campsite to some breathtakingly beautiful spots.

How to get there

By car: From Úbeda (N322, A315, A319), 2 km before Cazorla, take first right (after petrol station) and head for Quesada. Follow green signs to campsite. From Villacarillo (C323, JH3191, JV7101), as you reach Cazorla, turn right towards Peal de Becero (A319). Just before petrol station, turn left towards Quesada. Follow green signs to campsite. **Public transport:** Direct bus from Granada to Cazorla (2x daily). From Jaén and Úbeda there are various buses.

103. El Cortijo del Pino

James Connell & Antonia Ruano
Fernán Núñez 2,
18659 Albuñuelas, Granada.
Tel: 958-77 62 57 Fax: 958-77 63 50
cortijodelpino@eresmas.com
www.globalspirit.com/alpujarras/el_pino.htm
Open all year Language: E, GB, F
€ pn B&B: 1p 35; 2p 77.50; VAT 7%

House and surroundings

Hotel Cortijo de Pino is on a hillside above the village of Albuñuelas, looking out over the Lecrín valley, the Sierra Nevada and the Sierra de los Guajares. The old farmhouse has been converted into an intimate 5-room hotel. It has a lovely sitting room (with its original tile floor) and a small dining room. There is a large terrace shaded by a huge centenarian pine tree. Enjoy the beautiful gardens, swimming pool and patio where you can eat breakfast. There is also a studio: painting holidays are possible on request.

From the farm you can walk through the olive, almond, orange and lemon groves directly to the pine forest with its gurgling mountain streams. This part of the Alpujarras is famous for its picturesque white villages (on the Ruta de los Pueblos Blancos). Nearby attractions include Moorish architecture, the sea, Granada (45 km away) and the snow-capped mountains of the Sierra Nevada.

How to get there

Albuñuelas is situated about 40 km S of Granada. Owner provides information on public transport and directions to the house when you book your stay.

---

104. Cortijo La Chicharra & La Chaparra

Gerhard Schönhofer & Hilde Arnold
C/ Hondillo 102, 18420 Lanjarón, Granada
Tel/fax: 958-77 04 41 Mob: 657-33 39 13
chachibvista@wanadoo.es
www.ownersdirect.co.uk/Spain/s59.htm
Open all year Language: E, GB, D, I
Self-catering: €225(ls)-€450(hs) pw;
extra bed €9 pn

Farmhouses and surroundings

Cortijo La Chicharra (The Cricket) and La Chaparra (The Little Holly-Oak) are two traditional Alpujarran houses high up in the Sierra Nevada. This spot combines absolute peace and quiet with an incredible view! On a clear day you can see the Mediterranean.

The self-catering accommodation is very tastefully decorated. Both stone buildings have a large living room with a fireplace, kitchen, bathroom and 3 double bedrooms. Each has its own patio and large terrace. Next to La Chaparra there is a small pool of clean spring water. There is also a swimming pool on the premises. The farmhouses are in a beautiful garden with sweet chestnut trees. The owners tend a vegetable garden and the fruit and nut trees. They are also reforesting part of their property.

The farmhouses are a good starting point for walks in the Sierra Nevada nature reserve, mountain climbing, paragliding, riding and

bird watching. Other options include day trips to Granada (to Alhambra and gardens) or to the lovely fishing villages and beaches on the coast. Lanjarón, an old town known for its mineral water, is only 6.5 km away and has several restaurants that serve great local food.

How to get there

By car: Follow the owner from Lanjarón to the farm. **Public transport:** Train to Granada, bus to Lanjarón (Alsina-Graells bus departs 2x daily).

---

105. Cortijo La Lomilla

Marianne Vestjens & Harry van Meegen
Apartado 64, 18400 Orgiva, Granada
Tel: 958-34 70 80 / (040-212 95 08 in NL)
bdema@xs4all.nl
www.xs4all.nl/~bdema/lomilla/lomnl.html
Open all year Language: E, NL, GB, F
€ pn Camping: adult 4.80; tent to rent 5.15;
Self-catering: 185-210 pw; breakfast 6;
dinner 10.50; B!

Farm and surroundings

La Lomilla is a farm in Las Alpujarras, the mountain range between the snow-capped Sierra Nevada and the Mediterranean. The 3 ha of terraced farmland are planted with olive and fruit trees; there are also chickens on the premises. Guest can enjoy the great views, flowers, birds, butterflies and open spaces.

The terrain is somewhat difficult to negotiate, but the tent pitches are beautiful. Guests can also hire a small farmhouse with a fireplace and a fully equipped kitchen (sleeps 4), a *casi casita* (mini-house), and a big tent. Marianne bakes fresh bread and also prepares simple but tasty meals (on request) with her own home-grown organic produce.

Electricity on the farm comes from solar panels. The property is unlit at night, so be sure to bring your own flashlight. Children can play in the swimming pool, paddle pool and sandbox. Las Alpujarras and the Sierra Nevada are excellent for walking. You can take outstanding day trips or several-day hikes starting from La Lomilla or from one the characteristic white mountain villages of Pampaneira, Capileira, Pitres and Trévelez. The Dutch owners of the farm have marked out 10 walking trails, which vary in duration from 3 to 10 hrs. Málaga, Granada and Spain's southern coast are all easily within reach, even by bus.

How to get there

By car: From Orgiva, GR421 towards Pampaneira, Trevélez. 1 km after road to Soportújar and just before roadside chapel (Hermita del Padre Eterno) turn right on to unpaved road (downhill). Continue past quarry (approx. 200 m), then keep right (this part of road is also marked with red-and-white GR markings). Do not turn left into valley. After some 4.3 km, at white stone with red arrow, turn left to La Lomilla (another 700 m). Drive carefully (parts of road in poor condition). If you come by foot, there is a shorter route from Orgiva. **Public transport:** Train to Granada, then bus to Orgiva (Alsina-Graells bus, 2x daily). Also buses from Málaga or Almería to Orgiva. Call in advance to be picked up.

106. Cortijo La Torrera

Ana Losada Macías
Rambla de Lújar s/n, 18740 Castell de Ferro, Granada
Tel: 958-34 91 39 Mob: 655-11 30 16
analosada@navegalia.com
www.torrera.com
Open all year Language: E, GB
€ pn B&B: 2p 44; extra bed 9; Self-catering: 55; 375 pw; extra bed 9

Farm and surroundings

This traditional Alpujarran *cortijo* dates from the mid-19th century. The house sits on a hilltop with a view of the Mediterranean. It is surrounded by mountainous farmland. One ha is used to grow vegetables. The complex has been modernized and upgraded while retaining the traditional architecture and rustic feeling. In the main building there are 3 double bedrooms with their own bathrooms. There is also a small, self-catering apartment with a comfortable living room and a fireplace. The old barn has been converted into a bar/restaurant open at weekends, where the owners, an English-Spanish couple, serve traditional Andalusian meals. The *era*, the round threshing floor in front of the bar, is now used as a terrace. Outside the main entrance gate is a water basin in which you can swim. Various excursions are organized. The beaches of Castell de Ferro, a beautiful and relatively quiet resort, are just 3 km away. The Motril area (15 km away) is known for its sugar cane plantations and avocados. Restaurants in the fishing villages have a great selection of fresh seafood.

How to get there

By car: The owner will provide a map and give directions when you book your stay. **Public transport:** Train to Granada or Almería, then the bus to Castell de Ferro.

107. Cortijo Buena Vista

Juan Baena
Ctra de Mairena km 1, 18493 Laroles
Granada
Tel: 958-76 02 93 Mob: 679-87 27 05
cortijo@aldearural.com
www.aldearural.com/cortijo
Open all year Language: E, GB, F
€ pn Self-catering: 60-90(ls) 72-108(hs); extra bed 9.01; VAT 7%

ing signs to Cortijo Buena Vista. **Public transport:** Train to Granada or Gaudíx, then bus to Laroles.

108. Laveranda

Ismael Rodríguez
Fuentenueva s/n, 18858 Orce
Granada. Tel/fax: 958-34 43 80
laveranda@jazzfree.com
www.altipla.com/laveranda
Open all year Language: E, F
€ pn B&B: 2p 44(ls)-44.50(hs); Self-catering: 84.15-115.40; 420.70-577 pw; VAT 7%

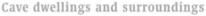

Farm and surroundings

Cortijo Buena Vista is situated in the Alpujarras mountains, in the middle of the Sierra Nevada nature reserve. This beautiful area bears the traces of many different civilizations. The farm owners tend 3 ha of fruit trees. Guests are welcome to help out with chores on the farm.

In the 19th century farmhouse, there is a group lodging and a self-catering apartment (max. 8 guests). There are several bedrooms, 2 living rooms and a fireplace. Meals are served on request. The host also organizes group activities such as archery, mountaineering for beginners, day trips along the GR-7 trail and educational field trips. You can also hire bicycles to explore the mountainous terrain.

The farm is an excellent base from which to explore the area's recreational and cultural attractions. The area's elevation varies incredibly, creating a diversity of ecosystems ranging from cold tundras to desert zones. The typical white villages feature examples of local architecture, for instance the narrow slate-covered chimneys. Local traditions and dishes reflect their Moorish roots.

How to get there

By car: From A92 Granada-Almería, head for Almería, then exit towards La Calahorra (A337). Continue for 30 km (past Puerto de la Ragua) until you get to Laroles. Upon entering village, turn right on to GR431 towards Válor and Mairena. After approx. 1 km, start follow-

Cave dwellings and surroundings

This 'cave hotel' consists of several dwellings built into the side of a mountain. In the 1970s, the abandoned cave village became popular among artists. Some 20 years ago, Ismael Rodríguez bought the homes and retreated from the world of conventional housing. He furnished the caves with beautiful antique objects. They became world famous after the 'Man of Orce' was found nearby.

Several caves are rented out. Some have windows with a view of outdoors. There are 3 bathrooms. Dinner can be enjoyed in front of the fireplace or outside on the terrace, which offers a breathtaking view. Besides the hotel rooms there are 3 self-catering cave dwellings for max. 10 guests, with a living room, fireplace, kitchenette and bathroom.

Sights worth seeing include the nearby Sierra de María and Castril (both 30 km away) and the historical towns of Huéscar and Baza. The cave dwellings are situated in an impressive mountainous area with a desert climate. Granada is about 1 hour's drive from Cúllar-Baza.

How to get there

By car: From the A92N Murcia-Granada, take the Cullar-Galera-Huéscar exit. Drive toward Galera-Huéscar on the A330. At km marker 16, turn right on to the SE 33 towards Orce. Continue through the village (on the SE 35) in the direction of María. After approx. 7 km you will see a sign to Fuentenueva and on the left side a small sign to Laveranda. Follow signs to Laveranda. **Public transport:** Train to Guadix, then the bus to Orce. From there, take a taxi.

109. Cortijo Aloe Vera

Lola Viudez Parra
Las Cañadas de Almajalejo,
04600 Huércal-Overa, Almería
Tel: 950-52 88 96
Open all year Language: E, GB, F, I
€ pn B&B: 2p 33; Self-catering: 42;
dinner 9.01; VAT 7%

Farm and surroundings

This traditional Almerian farm is a few km outside of Huércal-Overa, in the eastern part of Almería province. Although it has been completely renovated and modernized, the typical architectural style of this region has been preserved.

The owner serves home-made cheese and bread and pastries, freshly baked in a traditional oven. Meals are prepared with produce from the organic vegetable garden. There are dogs, cats, chickens and goats in the farmyard. There is 1 self-catering apartment for hire. It has a living room with a fireplace, kitchen, bathroom and bedrooms (max. 4 guests). Bed and breakfast is available in another part of the farm, where there are 4 double bedrooms (2 with

their own bathroom, 2 share a bathroom). There is a swimming pool and a playground.

The farm is set in a quiet rural area with beautiful panoramic views. The nearby town of Huércal-Overa has a long history. It was once inhabited by Phoenicians, Carthaginians, Romans and Moors.

How to get there

By car: From the N340/E15, take exit 533 towards Santa María de Nieva. Immediately after crossing the viaduct over the motorway, drive towards Almajalejo. After a few km you will reach a road sign reading 'Cortijo Las Cañas'. **Public transport:** Train from Alicante to Lorca, then the bus to Huércal-Overa. From there, take a taxi.

110. Complejo Rural La Hierbabuena

Gema Barreiro & Francisco López Barrios
Los Lobos, 04619 Cuevas de Almanzora, Almería. Tel/fax: 950-16 86 97
hierbabuena@a2000.es
Open all year Language: E, GB
Camping & lodging: prices on request

Farm and surroundings

La Hierbabuena is an organic olive farm situated just a stone's throw from the Mediterranean Sea and behind the protected Sierra Almagrera. The owner, a journalist from Madrid, started the farm in 1993, and was the first in the valley to turn to organic farming. The farm offers 3 types of accommodation: a campground with 20 pitches, 1 double and 1 triple bedroom in the cortijo itself, and 4 self-catering Finnish log cabins (for 4 guests each). There is a swimming pool and a ping-pong table on the premises. Guests can sign up for classes in Spanish language, culture, history, music and flamenco dancing.

Great beaches are only a 5 minute drive away. A little further south you will find the Parque Natural de Cabo de Gata-Nijar bird sanctuary, featuring flamingos, dunes, marshlands and the *palmita*, the only indigenous European palm tree. The owner also organizes excursions to the caves of Cuevas de Almanzora and the Paraje Natural Karst de Yesos de Sorbas.

How to get there
By car: On E15/N340 S (towards Almería), take exit 537 to Cuevas del Almanzora. Continue on A332 through village and towards Aguilas and Los Lobos. Just past point where a road on your right leads to La Mulería, you see sign reading 'camping a 400 m'. Further down that road, there is a sign to 'camping La Hierbabuena'. **Public transport:** Regular bus service from Murcia, Almería and Cartagena to Vera. Then bus to Aguilas (2x daily). Get off at Los Lobos bus stop, walk 1 km to farm.

111. Cortijo El Nacimiento

María Valdés García
04639 Turre, Almería. Tel: 950-52 80 90
Open all year Language: E, GB, F
€ pn B&B: 2p 33; extra bed 16.50; dinner 9; VAT 7%

House and surroundings
From the terrace outside of this traditional Almerian house there is an impressive view of the vicinity. The house is surrounded by a 60 ha property, used partly for growing fruit trees. María tends the organic vegetable plot and Adolfo prepares delicious vegetarian meals. You are more than welcome to assist in the garden. The house draws water from its own source. There are 5 double bedrooms, 3 bathrooms and a common living room with a piano. Breakfast and dinner are served in the dining room.

This place is ideal for relaxing, walking, reading and studying. The owner is happy to sug-

gest nearby activities. You can ride a horse or donkey and the beach is less than 30 minutes away by car. Swimming is also possible at a small lake closer by. The owner organizes group activities (walking, cycling, climbing and water sports). In the village of Sorbas (20 km away), which perches high up on a rock, you will find the ruins of a Moorish fort. The surroundings of Sorbas are famous for their many caves.

How to get there
By car: From Turre (between Almería and Lorca), AL150 towards Cortijo Grande. Pass low-lying area, drive uphill towards Sierra Cabrera complex. Where road stops ascending and makes a curve to the left, drive straight on to unpaved road (at Cortijo Nacimiento sign); this road all the way to car park. **Public transport:** Train to Almería, then ENATCAR bus via Vera and Mojácar to Turre (7 km from house). Arrange in advance to be picked up.

112. Finca El Rincon de Tablas

Ingeborg Wiegand & Johann Haberl
Apdo 32 Cortijo Grande, 04639 Turre, Almería. Tel: 950-52 88 03
asinus@larural.es
http://personales.larural.es/asinus
Open all year except Jul & Aug
Language: E, D, GB
Self-catering: €250-€300 pw

Farm and surroundings
The Sierra Cabrera, between Almería and Murcia, is the idyllic valley setting for El Rincón de Tablas, a secluded, almost self-supporting organic farm surrounded by nothing but nature,

the nearest settlement being 5 km away. On a clear day, unspoilt views take the eye as far as the impressive mountain ranges of Morocco. Ingeborg and Johann run their certified organic farm here, growing vegetables and owning ten Andalusian donkeys (an indigenous breed) and two mules.

Accommodation is in the guesthouse in four high standard double rooms with shower and toilet en-suite. Only natural materials have been used for construction, solar panels take care of heating and electricity supply, private springs supply fresh water, and breakfast is prepared with organic products from their kitchen garden. You can do your own cooking in the kitchen, and there is a community room. You can add a little extra to your stay by joining one of the organized walks on foot, donkey or by bike: single day walks or several day excursions. The donkeys have a peaceful character and are a pleasure to ride; Johann and Ingeborg have found that donkey rides are ideal for children, as well as for stressed-out adults! Also, discovering Andalusia on a donkey helps you to better perceive its culture and people. Other activities organized are Tai Chi, yoga and meditation.

Spring and autumn are the best times to visit, but flowering almond trees in February are also a delight to be enjoyed. Cool winter nights are cosy and comfy, with central heating and a wood-burning stove. The coastline of this part of Andalusia has an almost natural character still, with the renowned Moorish town of Mojacar 18 kms away. Further inland you will find the grandeur of Granada with its Moorish history. Johann and Ingeborg are very happy to help you plan your trips, and have plenty of information available.

How to get there

E15/N340 towards Almeria, exit to Turre/Mojacar. After 10 km, 2 km before Turre, there is a big gate indicating an entry to a golf course. Turn off here and follow track and signposts for El Rincón de Tablas.

113. Cortijo de Garrido

Richard Peelen & Hedwig Schouten
Uleila del Campo, 04270 Sorbas, Almería
Mob: 600-88 23 31 (023-531 12 99 in NL)
richardpeelen@hetnet.nl
home.hetnet.nl/~richpeelen
Language: E, NL, GB, F, D
€ *pppn Camping: adult 3; Lodging: 8; breakfast 4.50; dinner 7; B!*

Farm and surroundings

Situated in a tiny hamlet of just 4 farms, this small *cortijo* is being renovated and lovingly restored to its original state. Solar panels generate electricity and silently pump water from a well. The water is heated by a solar boiler. There is also a solar oven, which hostess Hedwig uses whenever possible. Her simple, delicious meals are made from a friendly neighbour's organic produce. The lodgings are still partly under construction. Two large bedrooms, an eat-in kitchen and a bathroom are complete. The once-abandoned almond and fig trees are tended again and the property is being reforested. Part of the almond and fig harvest is processed on the farm.

This is a blissfully quiet place, very suitable for walking, reading, studying, painting, helping out with renovation, gardening or just relaxing. Cortijo de Garrido is open for business in spring and autumn only (exact dates available on request). Bird watching and nature excursions can be arranged.

The farm is in a part of Andalucía largely undiscovered by tourists, and which is among the sunniest, driest areas in Europe. It is a 30-minute drive to the Desierto de Tabernas, Europe's only desert and a protected geological site. Cabo de Gata-Nijar nature reserve is 50 km away. The coastal park features small fishing villages and tourist resorts, secluded bays and a nudist beach. Just 10 km away from the farm is another park, where you can take guided cave tours in the geologically unique mountains. The nearby villages of Uleila del Campo and Sorbas have restaurants, a supermarket and a weekly open-air market.

How to get there

By car: On the E15 S, take exit 514 to Sorbas. Driving N on the E15, take exit 504 to Sorbas (AL 140). Drive through Sorbas, heading for Almería on the N340. After a few hundred m, immediately after a bridge, turn right (on to the AL 813) towards Uleila del Campo. Between km markers 7 and 8, turn right on to a dirt road. Continue on this road for 1.7 km, ignoring any roads leading off to the left or right. Drive through the dry riverbed and continue uphill again. Cross the neighbour's property and drive another few hundred m to the farm (impassable for caravans!). **Public transport:** Train to Almería, then the bus to Uleila del Campo or Sorbas. Call in advance to be picked up.

114. Cortijo Los Baños

Juan Segura Pérez
Cortijo Los Baños s/n
04210 Lucainena de las Torres, Almería
Tel: 950-26 81 36 Mob: 696-49 83 65
info@cortijo-al-hamam.com
www.cortijo-al-hamam.com
Open all year Language: E, GB
€ *pppn B&B: 18.03-24.05; extra bed 12.02; dormitory: 15.03; lunch & dinner 7.81*

Farm and surroundings

Cortijo Los Baños, on the northern slopes of the Alhamilla mountains, forms a green oasis in the desert-like landscape of Almería. The farm was built in the early 18th century at the site of public baths which were famous for their therapeutic effect. The buildings have been restored, and are now in use as a rural guest house specializing in health. It is a perfect place to seek rejuvenation, to breathe clean air, take a medicinal bath, sunbathe and relax. Vegetarian and vegan meals prepared with home-grown organic produce are served. There are 5 double rooms with bathrooms *en suite*, 2 double rooms that share a bathroom, and 2 dormitories for 8 people each. The dining room and lounge are cosy and comfortable. There is a swimming pool on the premises. You can choose from a variety of workshops: yoga, massage, health, creative subjects and organic farming. Guests can also sign up for guided cultural and nature excursions.

There are no less than 4 nature reserves in the area: Parajes Naturales Karst de Yesos de Sorbas (with subterranean caves), Desierto de Tabernas, Sierra de Alhamilla and the large Parque Natural de Cabo de Gata-Nijar. The latter three are all bird sanctuaries. Just 1 km from the farm is the village of Lucainena, with its charming white houses and mining history, including a small railway used to transport minerals to the coast.

How to get there

By car: On the E15 Murcia - Almería, take the A370 to Sorbas and El Barranco de los Lobos. Approx. 2 km after this village, turn left on to the AL102 to Lucainena de las Torres. **Public transport:** Information available on request.

115. Montaña de Firgas

Arnaldo Llerena Pinto
Montaña de Firgas 22
35430 Firgas, Gran Canaria
Tel: 961-60 01 78 Mob: 626-13 15 56
arnalben@teleline.es
Open all year Language: E, GB
€ *pp Self-catering: 13-15(ls) 16-20(hs) pn;*
91-105(ls) 112-140(hs) pw

House and surroundings

Montaña de Firgas (elev. 600 m) is just outside the village of Firgas, far from Gran Canaria's mass tourism. The patio offers a splendid view of the ocean. The house is situated between the narrow gorges of Tilos de Moya and Virgen Azuaje, on the edge of the Parque Rural de Dorama - famous for its *laurisilva*, forests that date back to the Tertiary. The house contains 2 spacious self-catering apartments: one with 2 double bedrooms (max. 4 guests) and one with 5 quadruple bedrooms, large enough for 20 people. A large garden full of fruit trees (apple, pear, Japanese persimmon and avocado) surrounds the house. The neighbour sells organic produce. The owner pioneers permaculture in the Canary Islands.

From the house, you can walk the *caminos reales*, old army routes to farms and villages in the vicinity or through the Virgen Azuaje gorge. Gran Canaria has over 240 km of beaches and 31 nature reserves. Beaches that are not totally overrun by tourists are Playa de Melenara and Playa de Salinetas near the village of Telde. The beautiful but very crowded Playa de Las Canteras is just north of Las Palmas. Good places to dive are Playa de Sardina near Galdar and Playa del Carbón. The cave near Galdar features beautiful geometric drawings. Firgas

is famous for its spring water. The village celebrates San Luis day on June 21 and San Roque day (the patron saint of Firgas) on August 16.

How to get there

By car: Drive from the airport to Las Palmas. On the Avenida Marítima, head for Arucas en Agaete (GC3 and C813, through the tunnel). In Arucas, turn left to La Caldera and then Firgas. In Firgas, look for Calle Calvario, nr. 46. This is where the road to Montaña de Firgas starts (500 m). **Public transport:** Bus from airport to Las Palmas bus station. There, bus to Firgas.

116. Atalaya de la Rosa del Taro

Silverio López Márquez
Atalaya de la Rosa del Taro 92
35600 Puerto del Rosario, Fuerteventura
Tel: 928-17 51 08 Fax: 928-53 26 76
Open all year Language: E, GB
€ *pn Self-catering: 42.07; extra bed 6.01*

House and surroundings

La Atalaya de la Rosa del Taro (elev. 367 m) is situated in central Fuerteventura, one of the Canary Islands. The north-eastern trade winds that carry desert sand from the West-African Sahara are also responsible for the warm weather, the desert-like landscape and the beautiful beaches. There are 2 self-catering apartments in 2 natural stone cabins. The cabins sleep 2 (plus 1 cot by arrangement). Each has its own kitchenette and a patio that offers peace, quiet and incredible views. Solar panels provide electricity, and water is recycled. There is a pottery and etching studio where you can take classes or work independently. Owner Silverio has created a bird feed which attracts a variety of birds including the occasional *Tara-billa Canaria (Saxicola dacotiae)*, a species unique to Fuerteventura. Free room and board for those participating in the *Territorio Imaginaria* (Imaginary Landscape) project, which involves reforestation using indigenous trees, shrubs and plants and a water purification scheme.

Fuerteventura has various cultural and natural monuments as well as protected landscapes and 3 nature reserves. The island is ideal for many outdoor sports such as walking, riding, diving, swimming, surfing and sailing. The little island of Los Lobos, north of Fuerteventura, is home to several trees and plants that grow nowhere else in the world.

How to get there

By car: From Puerto del Rosario, take the 610/FV20 towards Antigua. In Tesjuates, turn left on to the FV430 to Triquivijate. Four km down this road, at the sign 'cerámica/pottery', turn left on to an unpaved road and drive the remaining 1 km. **Public transport:** From Puerto del Rosario, take a bus or a taxi to the intersection in Triquivijate. From there, it is another 5 km: call in advance to be picked up.

Certified organic farm (see page 110)

Certification of sustainable and environment-friendly management by the Generalitat de Catalunya (see page 111)

Member of Nekazalturismoa, the Basque Farm Tourism organisation

Member of the Association of breeders of Xalda sheep (see pages 111 and 155)

Member of RAAR, Network of Andalusian Rural Tourism

In Spain ECEAT co-operates in some regions with local rural tourism and agricultural organizations

In Navarra

ITGA - Instituto Técnico y de Gestión Agraria de Navarra
Departmento de Formación Agraria
Avda. San Jorge 81, Entreplanta Dcha
31012 Pamplona, Navarra
Tel: 948 27 80 11 Fax: 948 25 13 21
itgasanjorge@sarenet.es

In Euskadi

NEKAZALTURISMOA

Nekazalturismoa
Done Mikel Auzoa, 11-1°.
48200 Garai, Biskaia, Euskadi
Tel: 94 620 11 88 / 94 620 11 63 Central Reservation desk
agroturismo@nekatur.net
www.nekatur.net

In Asturias

AGROTURISMO
en Asturias

Asociación de Agroturismo en Asturias
Tel: 985 89 05 50 / 985 89 05 26
Tel: 985 20 77 42 Central Reservation desk
info@agroturismo.net
www.agroturismo.net

In Andalucía

RAAR - Red Andaluza de Alojamientos Rurales
Apartado 2035
04080 Almería, Andalucía
Tel: 902 44 22 33 Central Reservation desk
Fax: 950 27 04 31 infoal@raar.es
www.raar.es

Spain

Farm nº 76. Son Mayol

Vegetarian and macrobiotic restaurants in Spain

Catalunya

Province of Girona
La Polenta
Cort. Reial, 6
17004 Girona
Tel: 972 20 93 74

Til La
Avda. Sant Narcis, 65
17005 Girona
Tel: 972 23 45 45

Province of Barcelona
Arco Iris
Roger de Flor, 216
Barcelona
Tel: 93 458 22 83

Biocenter
Pintor Fortuny, 25
08001 Barcelona
Tel: 93 301 45 83

Comme Bio
Restaurant & health food store
Gran Via, 603 / Via Laietana, 28
08003 Barcelona
Tel: 93 301 03 76 / 93 319 89 68

Restaurante Vegetariano Hindú Govinda
Pza. de la Villa de Madrid, 4B
08002 Barcelona
Tel: 93 318 77 29
mahaiana@seeker.es

Juicy Jones
C/Cardenal Casañas, 7
Barcelona
Tel: 93 302 43 30

L'Hortet
C/Pintor Fortuny, 32
Barcelona
Tel: 93 317 61 89

L'Illa de Gràcia
C/ Sant Doménec, 19
Barcelona
Tel: 93 238 02 29

La Buena Tierra
C/ Encarnación, 56
Barcelona
Tel: 93 219 82 13

La Granja Biológica
Sant Eusebi, 64
08006 Barcelona
Tel: 93 201 57 50

La Riera
Regent Mendieta, 15
Barcelona
Tel: 93 448 20 16

Maoz Falafel
Carrer de Ferran, 13
Barcelona

Restaurante Vegetariano Oriental
C/Industria, 132
Barcelona
Tel: 93 436 56 01

Self Naturista
C/Santa Anna, 11-17
Barcelona
Tel: 93 318 23 88

Province of Tarragona
El Vegetariano de Reus
Carnisserias Bellas, 8
43201 Reus
Tel: 977 34 14 08

Aragón

Province of Zaragoza
La Zanahoria
Tarragona, 4
50005 Zaragoza
Tel: 976 35 87 94

Navarra

Goizane
Irunlarrea, 68
31008 Pamplona
Tel: 948 26 39 27

Euskadi

Province of Gipuzkoa
Macrobiotika
Restaurant & health food store
Intxaurrondo, 52-54
Donostia (San Sebastián)
Tel: 943 27 04 89 / 943 27 66 38

Province of Araba
Rest. Vegetariano Museo del Órgano
Manuel Iradier, 80
01005 Vitoria
Tel: 945 26 40 48

Province of Biskaia
Garibolo
Fernández del Campo, 7
48010 Bilbao
Tel: 94 422 32 55

Ortua
Mazarredo, 18
Bilbao
Tel: 94 424 51 02

Restaurante vegetariano
Urkijo, 33
Bilbao
Tel: 94 444 55 98

Vegetariano
Gipuzkoa, 7 (Deusto)
Bilbao
Tel: 94 447 37 11

Armonia & Salud
Av. de Algorta
Getxo

La Rioja

Iruña
C/ Laurel, 8
Logroño
Tel: 941 22 00 64

Cantabria

Centro Macrobiótico Ignoramus
Alcázar de Toledo, 3
39008 Santander
Tel: 942 23 03 15

Asturias

Restaurante Byblos
Manuel Estrada, 1
Oviedo
Tel: 985 27 44 64

Galicia

Province of A Coruña
Bánia
C/ Cordelería, 7
A Coruña
Tel: 981 22 13 01

O Vexetariano
C/ Puerta de Aires, 3 bajo
15001 A Coruña
Tel: 981 21 38 26

Cabaliño do Demo
Rua Aller Ulloa, 7 bajo
Santiago de Campostela
Tel: 981 58 81 46

Province of Pontevedra
Gálgala
Placer, 4
36202 Pontevedra
Tel: 986 22 14 17

Sabor Sabor
Sta Clara, 33
36002 Pontevedra
Tel: 986 84 07 95

Cúrcuma
Brasil, 4
36204 Vigo
Tel: 986 41 11 27

Castilla y León

Province of Valladolid
El huerto de Melibea
Avenida de Santander, 45
Valladolid
Tel: 983 26 86 76

Province of Segovia
Almuzara
Marqués del Arco, 3
Segovia
Tel: 921 46 06 22

Province of Salamanca
El Grillo Azul
C/ El Grillo, 1
Salamanca
Tel: 932 21 92 33

Comunidad de Madrid
Andralama
Pº Infanta Isabel, 21
28014 Madrid
Tel: 91 501 70 13

Ceres
Topete, 32
Madrid
Tel: 91 553 77 28

Ecocentro Esquilache
C/ Esquilache, 4
28003 Madrid
Tel: 91 535 17 98 / 91 533 31 07

Estragon
Plaza De La Paja, 10
28005 Madrid
Tel: 91 365 89 82

El Grandero de Lavapies
C/ de Argumosa, 10
Madrid
Tel: 91 467 76 11

El Maná Ecobiólogica
C/ Hernani, 36-38
Madrid

El Restaurante Vegetariano
Marques de Santa Ana, 34
28004 Madrid
Tel: 91 532 09 27

Elqui
Calle Buena Vista, 18
Madrid
Tel: 91 4680462

Isla del Tesoro
Manuela Malasaña, 3
28004 Madrid
Tel: 91 593 14 40

La Biotika
Amor De Dios, 3
28014 Madrid
Tel: 91 429 07 80

La Galette
Conde de Aranda, 11 /
Barbara de Braganza, 10
Madrid
Tel: 91 576 06 41 / 91 319 31 48

La Granja
C/de San Andres, 11
Madrid
Tel: 91 532 87 93

La Mazorca
Paseo Infanta Isabel, 21
Madrid
Tel: 91 501 70 13

Restaurante Artemisa
Tres Cruces 4
28013 Madrid
Tel: 91 521 87 21

Restaurante Chez Pomme
C/ Pelayo, 4
28004 Madrid
Tel: 91 532 16 46
Restaurante La Naturel
C/ de Zorilla, 11
Madrid
Tel: 91 369 47 09

Solaire
Santa María de la Cabeza, 41
28045 Madrid
Tel: 91 468 23 04

Vivavegana
C/ de Pelayo
Madrid

Castilla-La Mancha

Province of Guadalajara
Amparito Roca
Toledo, 19
Guadalajara
Tel: 949 21 46 39

Comundidad Valenciana

Province of Valencia
Casa Vegetariana Salud
Conde Altea, 44, Bajo Izq.
46005 Valencia
Tel: 96 374 43 61

La Lluna
San Ramón, 23
46003 Valencia
Tel: 96 392 21 46

Restaurant Ana Eva
Turia, 49
Valencia
Tel: 96 331 53 69

Restaurant Les Maduixes
Daoiz y Velarde, 4
Valencia
Tel: 96 369 45 96

The Nature
C/ Ramon y Cajal, 36
Valencia
Tel: 96 394 01 41

Province of Alicante

L'Indret
Garcia Morato, 5
Alicante
Tel: 96 521 66 14

Restaurante Vegetariano
Pedro Lorca, 13
Alicante
Tel: 96 670 66 83

Caña de Azucar
Fora Mur, Bº 3-A
Denia

Comedor Vegetariano
Capitan Alfonso Vives, 11 A-C
03201 Elche
Tel: 96 546 46 95

Islas Baleares

Mallorca
Bon Lloc
Calle Sant Feliu, 7
Palma de Mallorca
Tel: 971 71 86 17

Julivert Restaurant Vegetaria
C/ Santiago Rusiñol, 13
Palma de Mallorca
Tel: 971 71 75 05

Murcia

El Girasol
C/ San José, 22
Murcia
Tel: 968 21 29 65

Maná
Pza. Junterones, 6
30004 Murcia
Tel: 968 28 58 24

Extremadura

Province of Cáceres
La Mandrágora
General Mardaño, 19
Cáceres

Province of Badajoz
La Ochava
Zurbarán, 15
Badajoz
Tel: 924 24 70 15

Andalucía

Province of Granada
L'Atelier
Vegan & Vegetarian Restau-
rant and Guesthouse
C/ Alberca, s/n
Mecina Fondales
Tel/Fax: 958 85 75 01
atelier@ivu.org
www.ivu.org/atelier

 *The Vegan and
Vegetarian Guide
to Restaurants in
Spain, with a hun-
dred listings, by
Jean Claude Jus-
ton, owner of L'Ateliers, is now
out at the price of €8 including postage (Europe only).
Available through the Atelier
website and through
www.vegetarianguides.com*

Raices
Pablo Picasso, 30
Granada
Tel: 958 12 01 03

Province of Málaga
Vegetalia
C/ Sta Isabel, 8
Los Boliches
29640 Fuengirola
Tel: 952 58 60 31

Cañadu
Plaza de la Merced, 21
29012 Málaga
Tel: 95 222 90 56

*El Vegetariano de San
Bernardo*
C/ Niño De Guevara, 4
29008 Málaga
Tel: 95 222 95 87

*El Vegetariano de La
Alcazabilla*
Pozo Del Rey, 5 B
29015 Málaga
Tel: 95 221 48 58

*Restaurante Asociación
Vegetariana*
Carretería, 82 1º
29008 Málaga
Tel: 952 40 00 52

Salomón
C/ Salomón, 8 B
29013 Málaga
Tel: 952 26 21 46

Espiga
Avenida Joan Miro, 16
Torremolinos
Tel: 952 05 21 02

Henleys
C/ Cristo
Nerja
Tel: 952 52 63 71

Relax
C/ Los Remedios, 27
Ronda
Tel: 952 87 72 07

Province of Sevilla
Jalea Real
Sor Angela De La Cruz,37
41003 Sevilla
Tel: 954 21 61 03

La Mandrágora
Albuera, 1
41001 Sevilla
Tel: 954 22 01 84

Zucchero
Golfo s/n
41003 Sevilla
Tel: 954 22 01 84

Ceuta y Melilla

Duala
Carretera Alfonso XIII, 63
Melilla
Tel: 952 67 36 29

Oasis
Monte Hacho, San Antonio,
s/n
Ceuta
Tel: 956 51 59 25

Islas Canarias

Tenerife
Restaurante Kimpira
San Vicente Ferrer, 5
Santa Cruz de Tenerife
Tel: 922 24 26 06

arm no 37. Baserri Arruan Haundi

Useful addresses in Spain

Tourist offices for Spain

The Netherlands
Spaans Bureau voor Vreemdelingenverkeer
Oficina Española de Turismo
Laan van Meerdervoort 8-A
2517 AJ Den Haag
The Netherlands
Tel: +31 (0)70 346 59 00
Fax: +31 (0)70 364 98 59
infolahaya@tourspain.es
www.spaansverkeersbureau.nl
Opening hours: 09:00-17:00 Mon. to Fri.

United Kingdom
Spanish Tourist Office
22-23 Manchester Square
London W1U 3PX
United Kingdom
Tel: +44 (0) 207 486 80 77
(09063 64 06 30 brochure request)
Fax: +44 (0) 207 486 80 34
londres@tourspain.es
www.tourspain.co.uk
Opening hours: 09:15 to 16:15, Mon. to Fri.

Organic Agriculture

Certifying organizations

Catalunya
Consejo Catalán de la Producción AE
Gran Vía de les Corts Catalanes, 612-614
08007 Barcelona
Tel: 93 304 67 00 Fax: 93 304 67 13
ccpae@correu.gencat.es

Aragón
Comité Aragones de Agricultura Ecológica
Instituto de Formación Agroambiental de
Movera
Chalet n° 1
50194 Zaragoza
Tel 976 58 69 04 Fax: 976 58 60 52
caaearagon@infonegocio.com

Navarra
Consejo de la Producción AE de Navarra
Avda. San Jorge, 81 - Entreplanta
31012 Pamplona
Tel/fax: 948 27 80 11 Fax: 948 25 13 21
itgsanjorge@sarenet.es

Euskadi
*Dirección de Política e Industria
Agroalimentaria*
Donostia (San Sebastián), 1
01010 Victoria-Gasteiz
Tel: 945 01 97 06 Fax: 945 01 97 01 02

La Rioja
*Dirección General de Investigación y
Desarrollo Rural*
Finca Valdegón Apdo 433
26080 Logroño
Tel: 941 29 11 50 Fax: 941 29 13 92
cida@larioja.org

Cantabria
Consejo Regulador de la AE de Cantabria
C/ Héroes Dos de Mayo, s/n
39600 Muriedas
Tel: 942 26 23 76 Fax 942 26 23 76
craecn@mundivia.es

Asturias
*Consejo de la producción Agraria Ecológica
del Principado de Asturias*
Avda. Prudencio González s/n
33424 Posada de Llanera
Tel/fax: 98 577 35 58

Galicia
Consejo Regulador de la AE de Galicia
Apdo. de correos 55
Rúa Pescaderías, 1
27400 Monforte de Lemos (Lugo)
Tel: 982 40 53 00 Fax: 982 41 65 30
craega@arrakis.es

Castilla y León
Consejo Regulador de AE de Castilla y León
C/Pío del Río Hortega, 1-5°A
47014 Valladolid
Tel: 983 34 38 55 / 983 34 26 40
Fax: 983 34 26 40
caecyl@memo.es
www.sister.es/castilla-leon/ecologica

Madrid
Comité de AE de la Comunidad de Madrid
C/ Bravo Murillo, 101
28020 Madrid
Tel: 686 41 13 86 Fax: 915 53 85 74
esmaae@teleline.es

Castilla-La Mancha
*Dirección General de Alimentación y
Cooperativas*
C/ Pintor Matías Moreno 4
45071 Toledo
Tel: 925 266751 Fax: 925 266722

Sohiscert, S.A.
C/ Hernan Pérez del Pulgar, 4 3°A
13001 Ciudad Real
Tel: 926 27 10 77 Fax: 926 27 10 78
sohiscert@sohiscert.com
www.sohiscert.com

Valencia
Comité de AE de la Comunidad Valenciana
Cami de la Marjal, s/n
46470 Albal (Valencia)
Tel: 961 22 05 60 Fax: 961 22 05 61
caecv@cae-cv.com

Baleares
Consejo Balear de la Producción Ecológica
C/Eusebi Estada, 145
07009 Palma de Mallorca
Tel/fax: 971177108
caeba@redestb.es

Murcia
Consejo de AE de la Región de Murcia
Avda. Río Segura, 7
30002 Murcia
Tel: 968 35 54 88 Fax: 968 22 33 07
caermurcia@teleline.es

Extremadura
*Consejo Regulador Agroalimentario
Ecológico de Extremadura*
C/ Padre Tomás, 4
06011 Badajoz
Tel: 924 21 50 66 Fax: 924 38 26 46

Andalucía
Comité Andaluz de Agricultura Ecológica
Apdo de correos 11107
Cortijo del Cuarto, s/n
Tel: 954 68 93 90 Fax: 954 68 04 35
caae@caae.es
www.caae.es

Sohiscert, S.A.
C/ Alcalde Fernández Heredia, 20
41710 Utrera (Sevilla)
Tel: 955 86 80 51 Fax: 955 86 81 37
sohiscert@sohiscert.com
www.sohiscert.com

Canarias
Consejo Regulador de la AE de Canarias
C/ Valentín Sanz, 4, 3°
38003 Santa Cruz de Tenerife
Tel: 922 24 62 80 Fax: 922 24 10 68

National associations for organic agriculture

Asociación Vida Sana
Clot, 39
08018 Barcelona (Catalunya)
Tel: 93 265 25 05 Fax: 93 265 24 45
info@vidasana.org
www.vidasana.org

*SEAE - Sociedad Española de Agricultura
Ecológica*
Granja la Peira
Apdo 107
46450 Benifaió (Valencia)
Tel: 961 78 80 60 Fax: 951 78 81 62
seaeseae@worldonline.es
www.agroecologia.net

Working on organic farms

Integral
Agricultura Ecologica
Apdo 2580
08080 Barcelona
Spain

WWOOF Independents
PO Box 2675
Lewes BN7 1RB
England, U.K.
www.wwoof.org
100 hosts listings in Spain

Environment & nature conservation

*Catalonian General Directorate of
Environmental Quality*
Diagonal, 523-525
08029 Barcelona
Tel.: 934 44 50 00
Fax: 934 19 76 30
wmadgqa@correu.gencat.es
www.gencat.es/mediamb/eng/qamb

*Asociación de Criadores de Oveja Xalda
Xalda-sheep Breeders Association*
Apdo 2117
33080 Oviedo, Asturias
Tel: 637 80 55 52
info@xalda.com
www.xalda.com/asociacion.htm

*Amigos de la Tierra
(Friends of the Earth-Spain)*
Avda. De Canillejas a Vicalvaro 82, 4°
28022 Madrid
Tel: 91 306 99 00 Fax: 91 313 48 93
tierra@tierra.org
www.tierra.org

Ecologistas en Acción
Marqués de Leganés, 12
28004 Madrid
Tel: 91 531 23 89 Fax: 91 531 26 11
www.ecologistasenaccion.org

Alentejo

Portugal

Flanking its much larger neighbour Spain on the western side of the Iberian Peninsula, Portugal stretches 560 km from north to south and about 200 km from the Spanish border to the Atlantic coast. Portugal also has offshore territories: the Azores and Madeira. On the mainland, the country has two distinct climates; wet, Atlantic air and heavy rainfall characterize the mountainous terrain of northern Portugal, while the south has a dry Mediterranean climate. Most large-scale tourism is concentrated in the south.

There are ten million Portuguese living in Portugal and three million who live and work abroad. The majority of the population is Roman Catholic. Portugal is a republic, divided into eighteen districts. Most inhabitants live in the coastal districts, which are also home to the country's industries. For years, these coastal areas have attracted many young people, a trend which has accelerated the drain on an already sparsely populated interior.

Portugal is the poorest member state in the European Union. It is sometimes compared to countries in Eastern Europe. Its main sources of income are the tourist industry in the south - Faro and the Algarve - and the money sent home by Portuguese working abroad. Industry and agriculture account for only a small slice of the GNP. Although the Carnation Revolution of 1974 brought badly needed educational reforms, 10 to 15 percent of the population is still illiterate.

In many people's minds, Portugal evokes images of sun, sea, blue skies, the Algarve, mules and carts laden with sun-ripened fruit. But this image belies the reality of millions of small-scale farmers who struggle to keep their local communities alive. Few foreigners know about the impoverished Portuguese interior. Portugal is witnessing a slow, but irreversible socio-economic destruction of rural life. As one farmer put it: "Only our big, silent rocks know the truth. If only they could talk."

Agriculture

Portuguese agriculture has always been dependent on local resources, knowledge, skills and institutions. Until recently, production was aimed at supplying the domestic market. The agricultural system was highly integrated and very diverse. However, this traditional system was disrupted by two factors: the state-run reforestation programmes and the exodus of the rural population. These influences upset the balance between agriculture and cattle-breeding, and between rural and economic development. These days, any hopes for healthy agricultural development are thwarted by the ongoing exodus, the growing dependence on external influences, the commercialization caused by Portugal's integration into international markets and the erosion of the socio-economic infrastructure. To make matters worse, yearly forest fires and increasing use of pesticides and fertilizers are impoverishing the soil. In recent decades, most European Union and state efforts have been aimed at agricultural specialization and modernization. A small group of wealthy farmers have profited from these schemes, but traditional farmers have only suffered. The reforms have also led to the waste of natural resources. Despite all this, the vast majority of Portuguese farmers still work by traditional methods. They can survive only if their livelihood is modernized sustainably, with respect for the local environment and social circumstances.

Organic farming

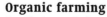

Certified organic farming - *Agricultura Biológica* - was introduced in 1985 by the Portuguese Organic Farming Association (AgroBio). Better access to agricultural subsidies has allowed the number of organic farmers to swell in recent years; there are now approximately 800 registered organic farmers. However, this growth is likely to be impeded by a lack of distribution and marketing structures. SoCert - which is affiliated with the French EcoCert organization - is the Portuguese body responsible for certification of organic products. Certification is so expensive that many farmers opt to sell their organic produce without a certificate. There are three consumer co-ops - BioCoop in Lisbon, NaturaCoop in Porto and Terra Preservada in Guarda - which guarantee quality products by creating a direct link from producer to consumer. These co-ops sell both certified and non-certified products.

Inspired partly by EU agricultural policy, there has recently been a shift towards the production of high-quality regional products. The challenge for the years to come is to create alternative markets for these traditional products, which include lamb, cheese, honey, preserves and smoked ham.

Interest in organic farming - which is an integral part of sustainable development - is clearly on the rise in Portugal. This is evident from the growing number of organizations active in rural development and environmental and wildlife protection. There is a strong call in rural areas for integrated regional development because this would encourage farmers to run environmentally friendly - but viable - farms and offer tourists a chance to enjoy the natural beauty and peacefulness of Portugal's farmland.

All the places to stay in this Portuguese guide not only meet the usual ECEAT standards, but also actively contribute to local development. Their owners believe that no sustainable development is possible without viable rural communities where the population can make a respectable living. All of the places to stay in this guide actively contribute in some way to the long-term sustainable development of rural Portugal.

Practical information about Portugal

Travelling to Portugal
By bus, via Paris - various bus companies, including Eurolines. By train, via Paris - TGV to Irún, regular train via Vilar Formosa, Guarda, Celorico da Beira, Gouveia, Coimbra, Lisbon or Oporto (transfer in Pampilhosa). By airplane on various airlines (ask travel agent for information).

Getting around in Portugal
Both train and bus are comfortable and inexpensive. It is easier to obtain information regarding the rail service than bus information. But travelling by bus is generally more comfortable, quicker and easier. When travelling from north to south, it is best to take direct bus lines. Bicycles can be carried along on the trains, although conductors may require you to pay a small fee. Taxis are relatively inexpensive.

Climate
Summers are generally pleasantly warm and dry, but very high temperatures can be expected in the southern provinces. Autumn and spring are the best times to visit these areas. Travellers are always advised to bring a warm jumper on trips to the mountains. In the mountains, winters are very cold, with rain and snow. On the coast, winters are mild, with temperatures generally above 15°C.

Judiaria Street in Évora

Eating and drinking

The Portuguese are real meat eaters. In rural areas the choice of vegetables is limited. Cabbage, onions and tomatoes are available at every farm. Vegetarians may have some difficulty finding a reasonable variety of food. Restaurants serve no vegetarian meals aside from cheese omelettes. And few farmers are eager to acquaint themselves with vegetarian cuisine!

In mountainous areas, drinking water from streams is usually better than bottled mineral water. In the cities, bottled water is recommended.

General ECEAT-Portugal camping prices

€2 small tent (2 persons)
€4 big tent (3 and more persons)
€4.50 caravan/camper van
€3 adult
€1.50 child
€2.car
€1 motor
€0.50 hot shower
€2 use of electricity
€1.50 pets

Maps

All 1:50,000 scale maps of Portugal are available in Lisbon at the Instituto Geografico e Cadastral (open Mon. through Fri. 9:00-11:00 a.m. and 1:00-4:30 p.m.). Take tram 28 from Chiado to the Basilica da Estrela. In Oporto, go to Editora, Praça Filipe de Lancastre (slightly more expensive than in Lisbon). Of course, you can also buy these maps at specialist travel bookstores in your home country.

The Beiras

Beira Interior

This area is dominated by the Serra da Estrela mountain range, part of the central Iberian range. The region is divided into the Beira Alta and the Beira Baixa. Bordered by the river Douro in the north and the river Tejo in the south, the Beira (edge) encompasses another three river valleys. The Mondego, Dão and Douro valleys are famous for their wines. The southern river valleys of the Beira Baixa also produce (citrus) fruit, vegetables and grains. The flat landscape looks much like that of Alentejo, although that region is almost used for growing grain. Compared to the coast, the Serra da Estrela (Star Ridge) is a relatively underdeveloped region. It has a distinct climate - hot and dry in summer, cold and wet in winter (when it is frequented for skiing). The mountain range is a huge massif and home to Portugal's highest summit (1,993 m). Clear traces of glacial erosion are one of the Serra da Estrela's most unique characteristics. Another is the vegetation; chestnut, oak and pine trees are indigenous to the area.

The terraced farmland is used as pasture and to grow potatoes, vegetables and corn. Herds of sheep and goats are kept in small fields, but are also allowed to graze on unfenced land with wild vegetation. The most famous product of this region is ewe's milk. This milk and a vegetable curdling agent called *cardo* are used to produce the famed Serra da Estrela cheese.

The desolate mountains of this region provide refuge to a wealth of fauna. Bird watchers may spot the falcon, eagle, buzzard, hoopoe, kite and the icterine warbler here. Other wildlife you are likely to encounter here include large lizards, snakes, otters, foxes and wild boar.

The entire region, once a no man's land between the Christians in the north and the Moors in the south, is one of great historical importance. On a visit you will find walled villages, mediaeval castles, baroque churches and chapels, Roman bridges and roads, Jewish synagogues and Celtic burial sites. The area has several interesting towns, such as Guarda - the highest-elevated and coldest town in Portugal - and the lovely little mountain town of Manteigas. There is also

The historical village of Piodão

Alcafache Spa, Mangualde

Belmonte, once home of the famous seafarer Pedro Alvares Cabral, and Monsanto, which is known as the 'quintessential Portuguese village'.

None of the economic troubles of this region deter the villagers from holding their various annual festivals, holy days and cultural festivities. Every village has its own summer festival, marked by the loud bang of fireworks. The people dedicate a day to each individual saint, and everyone takes the day off for an annual market.

The Serra da Estrela is the only region in Portugal with an extensive network of long-distance footpaths clearly marked on a walking map (scale 1:25,000). These routes are described in the booklet 'Discovering the Region of the Serra da Estrela', published by the Serra da Estrela National Park. Even seasoned hikers are advised to seek shelter and rest during the hottest hours of the day (11 a.m. to 4 p.m.) and always to bring along warm clothes.

The ECEAT accommodation in the Beira Interior varies widely, from very traditional sustainable sheep farms, to modern ecological farms, to small mixed farms.

Beira Litoral

This province differs considerably from the other Beira provinces. In the south, it borders on the Serra de Lousã, a verdant mountain range with a wide variety of flora and fauna, and 'forgotten' ancient villages. The environmental organization Quercus has set out a network of walking routes here and offers guided nature walks for groups. To the east, the province borders on the Serra da Estrela and the picturesque Serra de Açor, which has a centuries-old tradition of small-scale agriculture. In the heart of the province you will find the rolling dales of the Mondego valley. This section of the Mondego river winds from the Aguieira reservoir to Coimbra - a city which boasts an impressive university, two cathedrals and botanical gardens. Another town one should not miss is Conímbriga, once the most important Roman settlement in all of Portugal. North of Coimbra is the Parque Nacional do Buçaco, a national park of great cultural and historical importance with an tremendous diversity of indigenous and exotic vegetation. Beyond the park, the Mondego river runs through an area with intensive rice cultivation and into the sea at Figueira da Foz.

This part of the Atlantic coast, known as Costa da Prata (Silver Coast), is a very popular holiday spot for tourists from northern Portugal. But if you avoid the larger beach resorts, you will discover charming fishing villages and quiet beaches. Further north, the beautiful port of Aveiro lies on the Ria de Aveiro, a large inland lake in the delta of the Rio Vouga. The landscape of green meadows dissected by waterways is reminiscent of the Dutch polders. Even the bicycle, which is not exactly popular in Portugal, is commonplace here!

BEIR AMBIENTE

The BeirAmbiente Vila Soeiro co-ordination centre, 6300 Guarda, Tel/fax: 271 22 49 00, organizes walking and riding tours from farm to farm through the Serra da Estrela.

1. Quinta Casal da Fonte Grande

Adalberto Dias Lino & Maria do Carmo Rodrigues Santos Lino
Feital, 6420 Trancoso
Tel: 271-88 61 07
Open all year Language: PT, F, GB, D
€ pn Camping: tent 2-4; caravan 4.50; adult 2.50; child 1.25; Self-catering: 20-50

Farm and surroundings

This small organic farm serves as a model farm for the region. The young owners - a farmer, his wife and her brother - have established a foundation for wildlife protection. They organize small-scale cultural and environmental activities. Eight cows, 2 horses and a pony live on the farm. The owners grow potatoes, maize, beans, rye, sweet chestnuts, olives, figs, grapes and other fruit. They also grow vegetables and herbs in a garden at their home in the village. The owners sell vegetables, cheese, honey and their home-made sausage, marmalade and preserves. A large five-person house next to the farm can be rented.

The farm has traditional games for guests to play, donkeys are available for riding, and there are marked walks. It is a great environment for kids, with lots of animals (cats, dogs and ducks). Swimming is possible in a large water reservoir. If you wish, your hosts will accompany you on walks. There is a weekly open-air market nearby.

How to get there

By car: Quinta Casal da Fonte Grande is 30 km N of Guarda and 12 km SE of Trancoso. From Vila Franca das Naves head for Trancoso; after 4 km, turn right at the sign to Feital. In the village of Feital, you will find a small square with

a church. To the left is the farm owners' residence ('EL 1962'). Your host will accompany you to the farm 2 km down the road. **Public transport:** By train via Guarda to Vila Franca das Naves. From there, take a taxi or call for free pick-up service.

2. Quinta da Lameira

António Joaquim &
Maria Adelaide Campos
Rua do Eiró 16, Figueiró da Granja,
6370 Fornos de Algodres
Tel: 271-70 35 88
Open all year Language: PT, F
Self-catering: €40 pn

Farm and surroundings

This renovated farm set in the foothills of the Serra da Estrela is a perfect example of integrated rural development. The property has 3 ha of orchards (cherry and hazel trees) and 7 ha of reforested land. The owners keep 150 beehives and grow vegetables. They own 1 horse.

The farm is run by a friendly, vivacious couple who live in the neighbouring village, where they also manage a bar/shop offering local products such as honey, *chourizos* (sausage) and cheese. They have lodgings for hire in an attractively renovated traditional house (for 4 guests). The farmers are happy to accompany you on botanical or bird-watching excursions. The wooded hills abound with birds of prey. Other interesting sights include a necropolis, dolmen and a Neolithic fort.

How to get there

By car: The farm is 15 km W of Celorico da Beira and 7 km E of Fornos de Algodres. In

Figueiró da Granja, follow signs for Aguiar da Beira (EN 330); on the intersection, go straight, heading for Muxagata. At the next intersection, turn left. The farm is 300 m down this road. Follow signs for 'Alojamento Particular'. **Public transport:** Train and bus come to 3 km from the farm (crossroads Fornos-Figueiró). Trains run from Guarda and Coimbra (4x daily), buses from Guarda and Viseu (twice daily). Taxi from Fornos de Algodres.

How to get there

By car: Mizarela is 12 km W of Guarda. From Guarda take the N16 in the direction of Celorico da Beira. After approx. 16 km turn left towards Aldeia Viçosa. After 3 km, pass cross a bridge over the river. At the next crossing turn left towards Mizarela. After about 1 km, take the dirt road to the left, which leads to the farmhouse (after 100 m). **Public transport:** Train to Guarda-Gare station (20 km from the farm). Bus from Guarda to Aldeia Viçosa (bus stops 100 m from farm). Taxi from Guarda.

3. Quinta da Medroa

Francisco Viegas
Mizarela, 6300 Guarda
Tel: 271 59 81 53
Open all year Language: PT, F, GB
Camping: see page 115; B&B 2p €30 pn;
Self-catering: €50 pn B!

Farm and surroundings

This 7 ha farm is situated in the beautiful Mondego river valley. The river itself, lined by shady trees, forms one of the borders of the farm. This is an excellent holiday location for nature lovers, with magnificent surroundings for hikers, mountain climbers and anyone interested in botany, ornithology or geology. The huge granite farmhouse is divided into different apartments; they can take a total of 8 guests. Small-scale camping is also possible. The farm serves as a point of departure for walking trips down marked paths in the Serra da Estrela national park or visits to typical mountain villages such as Vila Soeiro, Videmonte or the famous Linhares.

The farm has a small swimming pool, and a reservoir (5 km away) offers swimming, canoeing and rowing.

4. Quinta do Carvalhal

Frits & Mirjam
Vila Soeiro, 6300 Guarda
Tel: 271-22 49 73; Mob: 914-79 07 02
bela.luz@clix.pt
Open May-Oct Language: PT, NL, GB, D, E
Camping: see page 115; caravan to rent
€10 pn

Farm and surroundings

This lovely small farm is situated near the village known as 'The End of the World'. The property consists of 0.5 ha of terraced farmland on the northern slopes of the Serra da Estrela. It is an ideal holiday spot for fun-loving nature lovers looking for peace, quiet and creativity. Managed by a young Dutch couple with 2 small children, the farm has an organic herb and vegetable garden and a variety of trees. An open kitchen offers convenient cooking facilities for guests. Two granite water tanks provide a chance to cool off. There are beautiful swimming spots in the Mondego river (1 km away). The farm is on two of the marked walking paths that cross the Serra da Estrela national park. Guests can take guided walking tours of 1 or more days in the direct surroundings and in other nature areas. Depending on the season, guests may help make herbal products, jellies, compotes, wine, brandy, cheese and bread. Fresh bread and a range of farm and village products can be purchased at the farm.

How to get there

By car: Vila Soeiro is 12 km W of Guarda. From Guarda take the N16 in the direction of Vale do Mondego. After about 6 km, turn left towards Barragem do Caldeirão. At the first crossing, turn right - down into the valley. Follow this winding road for about 3 km. After the Roman bridge, turn left and drive another 500 m. Once in the village of Vila Soeiro turn left and follow the dirt road. The farm is 500 m down this road, on the right-hand side. **Public transport:** Take the train to Guarda-Gare (20 km from the farm). Or take bus from Guarda to Aldeia Viçosa (bus stops 400 m from the farm). Taxi from Guarda €9. Pick-up service from Guarda €5.

---

5. Quinta do Ronfrio

Francisco Viegas & Maria Natividade Lopes
Trinta, 6300 Guarda
Tel: 271-59 81 53
Open all year Language: PT, F, GB
Camping: see page 115; Self-catering: €40 pn

Farm and surroundings

The very eco-minded middle-aged couple who run this traditional organic farm love to talk about the region's history. The farm has sheep, horses, chickens, ducks and a pig. From May to October, the sheep's milk is manually processed into Serra da Estrela cheese. Chestnut, hazel

and fruit trees can all be found on the farm, which borders on a large fresh water reservoir.

Campers can use the spacious kitchen with an open fireplace and refrigerator. There are also 2 small self-catering apartments for rent, large enough for 4 guests each. Breakfast products are included in the price of your stay. Home-made marmalade, compote and sausages are for sale.

Six km away, a river cuts through the mountainous landscape. Canoeing and swimming are permitted in the lake. Walking is recommended in Serra da Estrela national park. The pretty villages of Vila Soeiro and Videmonte are also worth visiting to see their Roman ruins.

How to get there

By car: Trinta is 15 km SW of Guarda. From Guarda, take the N16 towards Celorico. After 2 km, turn left towards Trinta (N338). Take this narrow road to Maçainhas de Baixo, where you cross the bridge over the lake. The farm is in the next sharp turn of the road. **Public transport:** Train to Guarda, then a JOALTO bus bound for Trinta; ask the driver to drop you off at Quinta do Ronfrio. Or taxi from Guarda.

---

6. Quinta da Alagoa

Geert & Ann
RN 232 n° 40, Vale de Amoreira,
6300 Guarda
Tel: 275-48 75 00
qtalagoa@hotmail.com
Open Apr-Oct Language: NL, F, GB
Camping (tents only): see page 115;
Self-catering: €8 pppn

Farm and surroundings

A Belgian couple with 2 children run this beautiful 5 ha family farm situated in the Serra da Estrela. It did not take them long to turn this isolated property into a flourishing organic farm. At the heart of the quinta is an old manor house built in slate and granite typical of this region. The property features olive and fruit orchards, vineyards and fields.

The farmhouse has a dormitory for max. 8 guests, open from Easter to October. The bathroom facilities are simple. The large, sunny terrace is ideal for enjoying a glass of house wine. Guests can play football, volleyball, badminton, etc. Swimming is permitted in the river that runs along the property, where you will also find a small sandy beach. The organic garden supplies ingredients for a healthy breakfast, lunch or dinner. You are welcome to help out with the production of wine, honey, preserves, etc. On request, the owners organize long-distance (several days) walking and cycling trips, minibus excursions and canoe trips.

Brochures with local information are available. The area is great for sunshine and relaxation, but there is also plenty to do. The surrounding mountains can be explored on foot or by (mountain) bike and there are local open-air markets and picturesque mountain villages to visit.

How to get there

By car: Vale de Amoreira is 25 km SW of Guarda and 10 km E of Manteigas. From Guarda, take the N232 to Manteigas and Vale de Amoreira. Just past Vale de Amoreira and just past (not across) the bridge, you will see the farm (a large slate building) to your right. From Manteigas, take the N232; the farm is 3 km past Sameiro, on the left-hand side. **Public transport:** Take a train to Guarda, then a bus to Vale de Amoreira. Paid pick-up service from Guarda.

7. Quinta das Poldras

Maddie van der Sande
Barrio Srª da Anunciação, São Romão
6270-596 Seia. Tel: 238-31 59 16
Mob: 918-26 04 33 Fax: 238-39 00 75
brinkesande@yahoo.com
Open all year Language: NL, GB, F, D
Self-catering pn: €35(ls)-€40(hs); B!

House and surroundings

You will find the 'Farm of the Polders' in the full glory of the Serra da Estrela national park, on a hillside adjacent to the river Alva. The renovated granite house has a perfect southern exposure and offers privacy despite its location near a village. The river with its shady trees offers clean, refreshing water, lovely spots to rest and playgrounds for children. This is an especially great place for bird watchers; several birds of prey - such as the eagle - are easy to spot. The spacious house has a living room with a fireplace, a kitchen, 3 bedrooms, a bathroom and terraces. Excellent for hikers in the summer and attractive in the winter as well, with the Serra da Estrela fully covered in snow.

How to get there

By car: São Romão is situated 3 km S of Seia. From Seia take road to São Romão. Here, follow road towards Torre via Srª do Desterro. After 2 km, just after sign reading 'Parque Natural da Serra da Estrela' turn right to Bairro Nª Srª da Anunciação. After 50 m, just before the village, take the small dirt road to the left. After 200 m you will arrive at the farmhouse. Caution: the road is quite steep. **Public transport:** Nelas train station is 27 km away. São Romão bus station is 2 km away. Taxi from S. Romão. Free pick-up service from S. Romão.

8. Quinta Covão de Santa Maria

António Gabriel Saraiva Direito
Covão de Santa Maria,
6260-150 Manteigas. Tel: 275-98 23 59
Open all year Language: PT, F
Camping (tents only): see page 115;
€ pn Lodging: 2p 30-32.50; dinner 10

Farm and surroundings
This big farm with a nostalgic feel is very re-motely situated in the middle of Serra da Es-trela national park. The beauty of the area is sometimes temporarily obscured from view by the mists that blanket the mountains at this high altitude. The property is bordered by a rivulet, which further downstream becomes the river Mondego. The farmers keep goats and use their milk to make organic cheese. There is a vegetable garden and a host of dif-ferent fruit trees. Beehives are kept for honey, and a watermill grinds the grain for your deli-cious home-baked bread. The family consists of a mother, 4 adult children and 1 small child.

The guest bedrooms and living room are to be found in the traditional, renovated farm-house and have separate access. There is a large kitchen for communal use. In winter, prices are slightly higher due to the cost of the central heating. The farm sells fruit, honey, eggs and meat.

The Serra da Estrela is home to many small mammals and rare plants. Manteigas with its trout ponds and Sabugueiro, the highest vil-lage in Portugal, are worth visiting. Swimming is possible in Vale de Rossim, 6 km away.

How to get there
By car: Manteigas is 45 km SE of Gouveia. On the main road from Gouveia to Manteigas (at approx. 15 km from Manteigas), turn on to the unpaved road across from hotel Pousada de São Lourenço. At the fork in the road (1 km), take the left road; the farm is 2 km further.
Public transport: Train to Gouveia, then a bus to Manteigas (get off at Pousada de São Lourenço). Or call to be picked up from Man-teigas (€5) or from Pousada de São Lourenço (free of charge).

---

9. Casa O Camponês

António Ferreira Machado
Rua do Caramelo, Nespereira,
6290 Gouveia. Tel: 238-49 46 46
Open all year Language: PT, E
Lodging pn: 2p €25

Farm and surroundings
The owner of Casa O Camponês, Mr Machado, has spent decades lobbying for healthy farm-ing in a clean landscape. He is an active mem-ber of the National Agricultural Federation and an organic farmer of regional products. De-spite his mature age, he restored a traditional bakery and renovated an old, granite house.

On the top floor of this house, he rents out 4 double bedrooms. The ground floor serves as a bar and exhibition centre for traditional agri-cultural tools, which helps him fund his cam-paign for healthy, local products. Most of the products Machado and his wife serve in the bar, such as wine, cheese, honey, raw ham and various types of sausages, are produced on the farm. All other products are locally produced.

Nespereira nestles in a green valley in the Serra da Estrela, where traditional farming is still practised. Historical villages like Vinhõ, Linhares and Arcozelo are not far away.

How to get there

By car: Nespereira is 3 km N of Gouveia. Coming from Coimbra (N17), at the intersection go straight in the direction of Gouveia. Take the first right to Nespereira. **Public transport:** Train to Nelas or Guarda (note: do NOT go to Gouveia station!). Then take a bus to Gouveia town and from there another bus to Nespereira.

10. Casa da Fonte

Clara dos Santos & Rui Lopes
Rua da Fonte 3, Arcozelo da Serra
6290 Gouveia
Tel: 238-77 45 62 Mob: 967-00 42 98
Open all year Language: PT, F, GB
Self-catering pppn: €12.50

Farm and surroundings

The farm is located in a small, traditional mountain hamlet, with narrow alleys and characteristic granite-built terraced houses. The tasteful dwelling has two rooms, a living with a sofabed, a kitchen with open fireplace and a bathroom. Owners Rui and Clara, who live in the village as well, are pioneers in changing the city for a life in the countryside. On their 2 ha farm, which is a ten minute walk from the village, they contribute to the renovation of sustainable agriculture. On various previously fallow fields, a wealth of organically grown ce-

reals, vegetables and herbs now prosper.

Arcozelo da Serra lies on the western slope of the Serra da Estrela, 3 km from the Mondego River. The surrounding area features curiously-shaped rock formations. For centuries, the terraces, now overgrown with pines and olive trees, were cultivated by farming communities. Clara, who is a painter of inner emotions, is greatly inspired by this ever-changing scenery.

How to get there

By car: The village of Arcozelo da Serra is located 7 km NW of Gouveia, close to the N17, between Coimbra and Celorico da Beira. From Coimbra go straight ahead at the crossing for Gouveia. First turning left towards Arcozelo. The house is 100 m from town square. Please ask for Rui and Clara. **Public transport:** Train to Mangualde (18 km). Bus to Gouveia (7 km). Taxi or paid pick-up service.

11. Quinta das Cegonhas

Rieke Marien & Gerard Duis
Nabaínhos, 6290-122 Melo
Tel: 238-74 58 86 cegonhas@cegonhas.com
www.cegonhas.com
Open Feb 15th-Jan 1st
Language: NL, D, GB, PT
€ pn: Camping: tent 2.50-3.30; caravan 3.30; camper van 4.60; adult 2.95; child 1.45; caravan to rent 25.95-35.95; breakfast 4.50; lunch pack 3; dinner 11

Farm and surroundings

The name *Quinta das Cegonhas* (House of the Storks) is taken from the 5 lanterns in the shape of stork's heads that give this fantastic manor house its distinctive look. A Dutch couple (with one child) run the 10 ha property, situated in a charming valley at the northern edge of the Serra da Estrela national park. High up on the ridges above this valley, you will find historical villages like Folgosinho and Linhares. The large campsite (4 ha) consists of terraced olive and fruit orchards. There are also several furnished caravans for rent (which sleep 2, 3 and 5). Guests can buy home-grown produce at the campsite. You can have a drink at the bar and take a dip in the swimming pool, both on the premises.

The immediate surroundings offer plenty of opportunities for walking, swimming and cycling. There are good itineraries and maps available for walking and cycling trips. The mountain village of Nabaínhos still maintains its traditional stonework, and continues other traditions of community life there.

How to get there
By car: Melo is 6 km N of Gouveia and 16 km S of Celorico da Beira. On the IP5 towards Viseu exit at junction 24 on to the N17 (Celorica da Beira - Coimbra). Go left at km 114 direction Nabaínhos/Melo and follow signs 'Quinta das Cegonhas'. **Public transport:** By train to Gouveia, then a bus or a taxi to the *quinta*.

12. Casa do Visconde

Lira & Pitum
Rua Arq. Keil do Amaral 82,
3525 Canas de Senhorim
Tel/fax: 232-67 10 23
Mob: 917-13 77 37 / 919-13 33 43
Open all year Language: PT, GB, F, D
€ pn Camping: tent 3-3.50; caravan 4; adult 3; child 1.50; B&B 2p 30-40; lunch & dinner 10; B!

House and surroundings
Casa do Visconde is a beautifully restored 17th century manor house in the historical centre of this lively village. It was part of the inheritance passed down by the Viscount of Pedralva, who served as minister of agriculture during the first republic (1910). The original character of his house was preserved in the renovation. The house still breathes an atmosphere of princes and fairies and offers lodging in 4 double

rooms (2 with private bathroom). Meals, with vegetarian options, are served with a nice glass of wine. Camping guests are asked to book two days before they plan to arrive.

The middle-aged couple who now occupy the house do some small farming: planting trees, growing berries and aromatic/medicinal herbs and raising chickens and ducks.

A park with very old trees and an olive orchard surrounds the house. Various buildings in the park are used as a venue for meetings, exhibitions and workshops on paper recycling, batik making, drawing, clay modelling, weaving, gold and silversmithing, drama and dance, and there is a swimming pool. Products for sale include herbs, jam, liqueur, dried flowers, cheese, honey and locally-produced artwork.

Casa do Visconde is situated in a densely-wooded area with many vineyards, between the Serra da Estrela and the Serra do Caramulo. This area is highly recommended for walking, riding and cycling. Thermal baths are just 5 km away.

How to get there
By car: Canas de Senhorim is situated on the N234, between Nelas and Carregal do Sal. Once you have reached the village, finding the house is easy: just across the street from the fire station. **Public transport:** Intercity train to Nelas or interregional to Canas-Felgueira and then take a bus or taxi (€5) to the village of Canas de Senhorim. Paid pick-up service from Nelas (€2)

13. Moinhos do Dão

Familie Van Dien
Tibaldinho, Alcafache, 3200 Viseu
Tel: 232-61 05 86 / (015-214 49 14 in NL)
Fax: 232-61 01 05
moinhos-info-nl@planet.nl
www.portugal-aktief.com
Open Apr-Oct Language: NL, GB, PT
€ pn Camping: tent 2.50-3.50; adult 2.50;
child 1.50; Self-catering pw: 175(ls)-245(hs);
dinner 13.60

Moinhos do Dão and surroundings

Moinhos do Dão (Mills on the Dão) campsite is tucked away on the beautiful banks of the river Dão just outside the small village of Tibaldinho. Until 1983, the water mills were used to grind rye and wheat. The miller and a shepherd were the valley's only remaining inhabitants. The shepherd is still there. It is possible to simply pitch a tent right on the banks of the river Dão or to rent accommodation ranging from a small caravan to a 4-sleeper cabin. Moinhos do Dão is simple but pleasing and convenient. It may not be the ideal place for those who seek luxury accommodation as it has no electricity and no indoor plumbing. On the other hand, it is just right for those who seek wonderfully candlelit evenings and fresh water from natural springs.

Guests have access to a washing room with solar-heated water, a restaurant and a telephone.

The river Dão is great for swimming and it has a small sandy beach on its banks. Walking itineraries and maps are available. You can take day trips or walks over several days, alone or with a guide. You can also take part in workshops such as painting, pottery and poetry.

How to get there

By car: The campsite is 10 km S of Viseu and 10 W of Mangualde. Leave the IP5 at Fagilde (between Mangualde and Viseu). Drive through Fagilde and go right into Villa Garcia. Turn right at the small upgoing road (Rua Escola), at the end turn left. Follow this road right through the village down to the convent. On this road follow the blue signs to Moinhos do Dão. Drive carefully: the road is bumpy.
Public transport: Train to Mangualde, then a taxi to the campsite. Limited pick-up service.

14. Quinta do Rio Dão

Annette Spork & Helmut Göbel
Canedo do Mato, 3530-106 Mangualde
Tel: 232-61 10 87 www.quintadoriodao.com
Open Jan 4th-Dec 12th; camping Easter-Oct
Language: PT, D, GB, F
€ pn Camping: tent 1.50-2.50; caravan 3;
adult 2; child 1; B&B 2p 45-60;
Self-catering: 25(ls)-40(hs)

Farm and surroundings

A young German couple with 2 children run this 17 ha organic farm on an idyllic spot near a creek in the river Dão. The couple grow fruit and vegetables and raise goats, sheep and chickens. They also have dogs, cats and a donkey.

Tents and small caravans are welcome on the campsite. Campers can use the solar-powered bathrooms/showers which are in a small patch of woodland. For a small fee, you may use the kitchen and get whatever produce you wish to buy from the organic vegetable garden. There is also a restored cottage, which accommodates 2 adults and 1 child.

The creek at 500 m, is ideal for swimming, canoeing (canoes for rent) and fishing. You can also take donkey-and-wagon rides through the surrounding meadows and woods, and visit the open-air markets in Penalva and Viseu. The owners also organize a 2-day mini-course in bread-baking, entitled 'From Grain to Bread'.

How to get there

By car: Mangualde is 15 km E of Viseu. From Viseu or Guarda on the IP5 (E80), take exit 19 towards Vila Meã. After a few metres, turn right at the sign to Barragem de Fragilde. Just before the reservoir (barragem), turn right on to an unpaved, winding road that leads to the farm (3 km). **Public transport:** Train to Mangualde station. From there, bus or taxi, or from the centre of Mangualde, taxi to Canedo do Chão. From the bar/minimart next to the church, there is a signposted route to the farm. The German bar owner is glad to help. From here, a jeep or donkey-drawn wagon can also pick you up. This service is free if you stay for a week or more.

15. Quinta da Chave Grande

Alberto Ymker & Gea Ymker-Kruise
Casfreires, 3560 Sátão
Tel: 232-66 55 52 bert@mail.telepac.pt
www.chave-grande.com
Open Apr-Nov Language: NL, GB, D
€ pn Camping: tent 3; caravan 3; camper van 5; adult 3; child 1.80; B&B 2p 19.50-32.50

Campsite and surroundings

Quinta da Chave Grande is a lively, spacious and comfortable campsite run by a Dutch couple. The property has 10 ha of sunny terraced slopes featuring grapevines, fig, chestnut, hazel and fruit trees. There are some caravans for rent and a double room.

There is a swimming pool, a nice bar to enjoy a good Portuguese wine, a tennis court and *jeu de boules* on the premises. Guests can take painting lessons and join small groups on excursions to nearby sights.

Do visit the nearby typical Portuguese villages; traditional festivities are held in all seasons, with the *vindimia* (the grape harvest) as a highlight. Attractions nearby include quarries, Neolithic burial chambers and the Piave, the cleanest river in Europe.

How to get there

By car: Sátão is 30 km NE of Viseu and 6 m SE of Vila Nova de Paiva. On the IP5 near Viseu, exit for Sátão. Before reaching Sátão you will start to see signs for the campsite. **Public transport:** Train to Mangualde, then a bus (bus stop across from campsite entrance) or a taxi. Pick-up service from Mangualde (€7.50).

16. Quinta da Comenda

Maria Laura & José Cardoso Rocha
3660-404 São Pedro do Sul
Tel: 226-17 98 89 / 232-71 11 01
Mob: 966-70 19 96 Fax: 226-18 34 91
quintadacomenda@hotmail.com
Open all year Language: PT, GB
B&B pn: 2p €75

Farm and surroundings

A farm with a long history, Quinta da Comenda was once owned by the mother of Portugal's first king. It is situated in the rich agricultural region of Lafões, in the Vouga river valley.

A Vouga tributary with its Roman bridge forms the southern border of the property. The current owners have struck a perfect balance between old traditions and modern technology to create a model of ecologically and economically sound farming. The property has vineyards (22 ha), apple orchards (10 ha) and woods (6 ha). There are 6 luxury double bedrooms for hire. The farm has a swimming pool, a children's play area, a chapel and a dining room. You can visit the wine cellar and water mill, and wander through the vineyards and orchards.

Only 2 km away there is a spa known throughout Portugal for its healing properties. The surroundings offer a wide selection of cultural and historical sights.

How to get there

By car: Quinta da Comenda is 2.5 km from S. Pedro do Sul and 15 km from Viseu. From Viseu, take the N16. At the Quinta da Comenda bus stop, you are already on the farm property. **Public transport:** Train to Mangualde, then a bus to Viseu or S. Pedro do Sul, or take a taxi.

17. Parque de Campismo Natural da Fraguinha

Maria de Lurdes & Eduardo Basso
Coelheira - Candal, 3660-043 S. Pedro do Sul
Tel/fax: 232-79 05 76
tantaserra@mail.telepac.pt
www.fraguinha.com
Open all year Language: PT, GB, F
€ pn Camping: tent 2.25-3.70; camper van 3.70; adult 2.25; child 1.25;
Self-catering: 37.40-62.35; extra bed 9; B!

Campsite and surroundings

At an altitude of 1,070 m in the heart of Maciço da Gralheira nature reserve, the 'Fraguinha' campsite seems to sit on top of the world. This forest, which has a brook running through it, was the site of a park until it was decommissioned in 1974; it was revived only a few years ago. The existing houses have been carefully converted into holiday apartments. A stone house with a schist roof houses a beautifully designed bar and restaurant. The region is known for its excellent meat products and aro-

matic herbs. The surrounding area abounds with Roman ruins, historical villages and natural beauty. The campsite is full of hidden secrets and stories waiting to be discovered by children and adults. Special activities are also organized for children, such as a treasure hunt and kite-making.

How to get there

By car: The village of Coelheira is 23 km NW of Viseu. From S. Pedro do Sul take the N 227 for approx. 4 km. Then turn right towards Sá. Note the sign indicating 'Fraguinha'. After about 900 m turn right towards Sá; 2 km after the sign reading 'Fraguinha', turn right again. From here, you must take the steep mountain road (10 km). At the next sign to 'Fraguinha' turn left towards Candal and Arouca. After passing Coelheira, you will see a road off to your left with a sign to the campsite. **Public transport:** Take a bus to S. Pedro do Sul. The bus station is 23 km from the campsite. Free pick-up service from S. Pedro do Sul.

18. Quinta da Cerca

Leonardo Min (Dardo) &
Beatrys Franssens (Bie)
Midões, Casal da Senhora, 3420 Tábua
Tel: 235-46 42 36 (02-262 00 34 in B)
mindardo@netc.pt
Open Jan 15th-Dec 15th Language: PT, GB, F, NL
€ pn Camping: tent 2.75-5; caravan 8; adult 3.50; caravan to rent 20-37; Lodging: 2p 25-30; breakfast 3.75; lunch 6.25; dinner 9.50

Farm and surroundings

Quinta da Cerca (15 ha) sits on a hilltop with a view of the mountains. The renovated granite manor is the heart of this place, and offers a pleasant stay for those who value hospitality and meeting others above luxury. The organic vegetable gardens assure healthy (vegetarian) meals, with dishes from all over the world. The sun terrace, pleasant communal room and shaded dance floor near the river can be used by those who want to organise a course. Next to the house and its guest rooms (3 double rooms and one for 4 persons) there is an open-air bar (summers only) and a swimming pool. Under the eucalyptus trees on the slope there are 6 caravans for hire, and you can pitch your tent in the olive orchard. Rye bread, honey, olive oil, jam and vegetables are for sale.

A small river cuts across the property. In summertime, dardo takes guests on caving expeditions upstream. Explore the wide variety of walking routes in the surrounding mountains which start at the quinta. Folder on request.

How to get there

By car: Quinta da Cerca is located between Oliveira do Hospital and Tábua. From Tábua, take the new road to Midões. After Midões (8 km), head for Casal da Senhora, until you reach the 2nd sheltered bus stop. Turn right (uphill) and follow the signs 'Quinta da Cerca' to the farm. From Oliveira do Hospital, take the road to Tábua until the first large fork after 2 km. Turn right towards Travanca de Lagos. Via Travanca, you will reach Midões. Casal da Senhora is 1 km further. **Public transport:** By train to Carregal do Sal or Santa Comba Dão, then by bus to Casal da Senhora.

19. Quinta do Vale da Cabra

Philipe Ghesquire & Patricia Dupont
Meruge, 3400 Oliveira do Hospital
Tel: 238-60 20 96
patricia.dupont@oninet.pt
Open all year Language: PT, B, GB, F, D
€ pn Camping: see page 115; Self-catering pw: 300; breakfast 4; dinner 9.75

Farm and surroundings

Picture this: eucalyptus, pines, cork oaks and lavender on the south face of a mountain overlooking the Serra da Estrela. Look more closely and see vineyards, olive and fig trees, a man with a packed donkey, a woman proudly carrying a basket of the day's harvest on her head. Now listen, and breathe in. Silence and the smell of herbs and flowers. This beautiful setting is where Patricia and Philipe established their campsite.

The 8 ha site has spacious and shady pitches, a swimming pool, a paddling pool, a bar and a terrace. Children can make friends with the ducks, geese, dogs and the donkey. There are also several caravans for hire (which sleep 6, 4 and 3 guests) and one chalet.

Local specialities feature in the 3-course meal served in the evening. Sandwiches and home-made wine, jams, olives and cheese are also for sale. Itineraries and maps are available for those who wish to walk or cycle in the vicinity.

Nearby Caldas da Felgueira has thermal baths and the imposing Serra da Estrela is just a stone's throw away.

How to get there

By car: Meruge is between Seia and Oliveira do Hospital. In Oliveira do Hospital, follow signs to

Caldas da Felgueria. At intersection, turn right towards Sta Eulalia, then follow signs to campsite. From Seia, take N17, then take second road to the right. Continue through Folgosa da Madalena and Sameice, to centre of Sta Eulalia. Turn right and follow 'Camping' signs for 2 km. **Public transport:** Train to Nelas, then bus to Meruge. Buses or taxi from Oliveira do Hospital and Seia. Paid pick-up service from Meruge.

20. Quinta das Mestras

Leondra Wesdorp & Rob Horree
Nogueira do Cravo, 3400 Oliveira do Hospital
Tel: 238-60 29 88 Fax: 238-60 29 89
wesdorp@pandaline.pt
www.pandaline.pt/qtamestras
Open all year Language: GB, F, D, NL, PT
B&B pn: 2p €37.50-€42.50

Farm and surroundings

Situated in a small valley, the 7.5 ha property of Quinta das Mestras has terraced olive and fruit orchards surrounded by a thick pine forest through which a brook runs. The farm had been abandoned for years before it was restored. In 1995, an Australian translator named Leondra and a Dutch graphic artist named Rob decided to roll up their sleeves and do their part to repopulate the Portuguese countryside. They renovated the 2 granite houses and improved the soil conditions.

Currently, the hosts have a bed & breakfast in 1 of the houses, where they rent out 1 single and 3 double bedrooms. Higher up the hill, there are also 2 small houses which sleep 2 persons each.

The area is excellent for long walks in the mountains. Guests can also acquaint themselves with local agricultural practices, or visit archaeological ruins or the wine cellars where the famous Dão wines are stored.

How to get there

By car: Nogueira do Cravo is halfway between Guarda and Coimbra. In Oliveira do Hospital, head for Venda de Galizes (N17), then turn right towards Nogueira do Cravo and through the centre towards Bobadela. After 2 km Quinta das Mestras is on right-hand side (ochre-coloured house, blue mailbox). **Public transport:** Train to Santa Comba Dão, then bus to Oliveira do Hospital or Venda de Galizes; from there, taxi or call to be picked up (no charge).

21. Residencial o Miradouro

Marianne Jansen & Helder Pachorro Correia
Avô, 3400 Oliveira do Hospital
Tel: 238-67 73 46 / 67 61 91
Open all year except Jul 31st-Aug 15th
Language: PT, GB, NL
Lodging pn: 1p €12.50; 2p €22.50-€25

Pension and surroundings

This small pension in the picturesque village of Avô is a rustic, wood-furnished building which includes a restaurant (open June 1-Sept. 15) and a café that serves meals in the winter. Vegetarian meals are prepared on request. The owners organize (on request) various day trips, for example to an organic brandy distillery. Wine, brandy, sweet chestnuts, vegetables and jam are for sale.

The area is great for walking. Avô and Pomares have little sandy beaches along the river where you can sunbathe and swim. The historical town of Piodão is a point of interest.

How to get there

By car: Avô is 35 km SE of Santa Comba Dão and 10 km S of Venda de Galizes. In Avô, head for Pomares. After 1 km, past last houses of the village, turn left on to dirt road. Continue up this road for 1 km. Farm is 2nd residence you come across. **Public transport:** Train to Santa Comba Dão. Then bus to Avô. From Avô, taxi. Pick-up service from Venda de Galizes (€5).

---

22. Casa da Eira

Fernando & Lucinda Ventura Martins
Largo Dr. Pires de Carvalho, Casal de Ermio,
3200 Lousã. Tel: 239-99 17 12 / 239-99 18 60
Fax: 239-99 17 12
Open all year Language: PT, GB, F
B&B pn: 2p €35(ls)-€40(hs)

Bed & breakfast and surroundings

Casa da Eira is in the centre of Casal de Ermio village, facing the green flanks of the hills. The river Ceira courses through the valley below. Some nights the moon shines so brightly that electric lights are not needed. Besides the continuously murmuring river, the silence is broken only by the sound of ravens, cicadas, frogs and other animals.

In the house there are 3 double bedrooms with private bathrooms for rent. The owner is a beekeeper whose honey has received several European awards. The honey is part of a range of organic products sold in Fernando's shop in Lousã. Fernando teaches beekeeping and gladly shows his guests how the honey is produced.

The mountains of Lousã are covered with a great variety of trees. The most highly recommended activities there are swimming, fishing and walking. You can visit a mysterious medi-

aeval castle and various remote villages. Detailed itineraries, maps and other information are available from the Quercus environmental organization.

How to get there

By car: Situated 9 km N of Lousã and 40 km SE of Coimbra. N17 from Guarda to Coimbra. Pass Vila Nova de Poiares and turn left towards Lousã at next intersection. At next crossroads turn left again, towards Casal de Ermio (2.5 km down the road). Casa da Eira is on main square on right-hand side. **Public transport:** Train to Lousã, then bus or taxi to Casal de Ermio.

---

23. Casa da Cerdeira

Kerstin Thomas & Bernhard Axel Langer
Cerdeira, 3200 Lousã. Tel: 239-99 46 21
Open all year Language: PT, D, GB, F
Self-catering pn: €25

House and surroundings

After decades of neglect, the village of Cerdeira, situated in the Lousã mountains, is now experiencing a revival. The complex of some 25 slate houses, sheds and watermills, criss-crossed by narrow winding alleys, seems as if it grew naturally out of its surroundings. An action group was initiated to represent the interests of this beautiful piece of architectural heritage. Cerdeira has no local shops or cafés for its few permanent inhabitants.

Kirsten and Bernhard live here with their two children. They make honey and grapefruit juice, grow olives, and are also dedicated to making dolls and sculptures. They carefully restored the traditional cottage, which sleeps 4, introducing a tasteful mix of colours and light. Courses and workshops teaching sculpture and Portuguese are available by arrangement. Maps and information are available.

The partly cultivated surrounding landscape features pines, oak woods, chestnut trees and shrubs. This exceptionally beautiful and varied environment is enhanced by numerous little streams and rivers. Nearby Lousã is a lively and modern town offering a choice of events.

How to get there

By car: Cerdeira is 10 km S of Lousã and 40 km SE of Coimbra. At Lousã follow signs to Castanheira de Pêra passing Alfocheira, going uphill. After 9 km turn off at the sign for Cerdeira (1.5 km). **Public transport:** Train or bus to Lousã (10 km), then taxi or arrange pick-up (for payment).

Farms in the Vale de Ceira

Farms nrs 24, 25 and 26 follow the trails of River Ceira upstream from Góis, and are at 7 to 25 km distance from each other. Some farms border the river banks; others overlook it from high up in the mountains. Luggage transport can be arranged if you want to trek from one to the other. At Góis, capital of the Ceira Valley, there is a charming little market on Tuesdays; in mid-July a week full of cultural events is organised; around the ninth of August they have the annual farmer's fair; and each nearby village has its traditional summer festivities!

The area has a high density of foreign farmers, which is hardly surprising considering that during the past fifty years the majority of the locals of this poor area tried to find a better life in the cities and even abroad. The remaining Portuguese mainly live in the villages nowadays.

Arganil and Góis are approximately 50 km E of Coimbra and 30 km NE of Lousã.

24. Quinta Vale de Asna

Io Laubenthal & Jan Vollenhoven
Vale de Asna, Saião, 3330 Góis
Tel: 235-76 13 65
Open all year Language: PT, NL, D, GB
Camping (tents only): see page 115;
Self-catering pn: €20

Farm and surroundings

This remote farm (elev. 600 m) is on the southern flank of one of the western foothills of the Serra da Estrela. Every day the farmers herd their goats-which they keep for milk, cheese and meat-in the surrounding hills.

The tent pitches are among the olive trees, close to the farmhouse. The owners also rent out a small, gas-lit house, for a minimum of 2 nights. The bathroom is 200 m uphill, near the main house. You can bathe in the pure spring water or in the waterfall in the river 4 km a way. Once or twice a week you can enjoy a vegetarian meal prepared by the hosts. Cheese, milk and bread are for sale on the farm. You are welcome to help herd the goats and bake bread.

There are plenty of paths for (long distance) walks on the mountain ridge, along the river or to one of the picturesque villages in the area. Itineraries are available.

How to get there

By car: Vale de Asna is 45 km E of Coimbra. Halfway down the N342 from Arganil to Góis, at chapel, turn off towards Celavisa, Colmeal. Stay on this road (which becomes dirt road at Pracerias) and go uphill for 5 km until you reach clump of high pine trees and little pillar with water tap. There, turn left towards Colmeal. After 1.5 km, on right-hand side and going downhill, you see first the goat shed and then farmhouse. **Public transport:** Train to Coimbra, then bus to Arganil. Here, taxi or call to be picked up from Celavisa (for a fee).

25. Quinta do Ribeiro

Francien Nijhof & Klaas Zwart
Cepos, 3300 Arganil
Tel: 235-75 13 43
nijhofzwart@mail.telepac.pt
Open all year Language: NL, GB, D
Camping (tents only): see page 115;
Self-catering pn: €20; dinner €8

Farm and surroundings

This 6 ha organic farm is run by a young Dutch couple. It is surrounded by great natural beauty and is located just a few hundred metres from a river. Francien and Klaas grow vegetables and raise chickens and hogs. The campsite has 6 pitches. There is also a tiny renovated house which serves as a self-catering apartment for 2. Dinner is served on request.

The terraced landscape with its olive trees and springs offers enthusiastic walkers endless possibilities.

How to get there

By car: The farm is some 40 km E of Coimbra. From Arganil, head for Folques, then Torrozelas and from there towards Cepos. Four hundred m before Cepos, turn left on to a dirt road. Turn right on to another unpaved road and continue downhill for the next 3 km. **Public transport:** Train to Coimbra, then a bus to Arganil and another bus or a taxi to Cepos. Call to be picked up in Cepos.

26. Quinta Regada

Helga Asal & Manfred Hilscher
Cepos, 3300 Arganil
Tel: 235-76 13 37
Open Apr-Oct Language: D, GB
Camping (tents only): see page 115;
Self-catering pn: €20

Farm and surroundings

Helga and Manfred live together with their four children on this 10 hectare farm, peacefully nestling on the slopes which descend to the Ceira river. The surounding scenery includes wild overgrowth, herb gardens and undulating fields. Farm animals include goats, hens, a pig, rabbits, a mule, turtles, fish, dogs and cats. They sell various organic products, and vegetarian diets are catered for. There is plenty of space to pitch a tent and a newly built house for 4 persons with a kitchen and living-room, using solar energy. There are opportunities to walk, fish, swim and play in the small river Ceira and at other river beaches, and trips to the beautiful market town of Arganil.

How to get there

By car: From Arganil towards Folques, then towards Colmeal passing Torrozelas and Cepos. After 2 km turn left on an unpaved track. Farm is at the end. **Public transport:** Train to Coimbra (50 km). Bus from Arganil to Cepos. Taxi from Arganil or pick-up service from Cepos (by arrangement).

27. Quinta das Bouchas

António Leal de Carvalho &
Maria de Lurdes Presunto Leal
Taliscas, Paúl
6215 Unhais da Serra
Tel: 275-96 14 11
Open Apr-Nov Language: PT, GB
Camping: see page 115;
caravan to rent €17.50 pn

Campsite and surroundings

Quinta das Bouchas is a 30 ha traditional mixed farm with cows, grain fields and olive and fruit trees. The host family (a farmer, his wife and 2 children) are very hospitable. The 12 pitch campsite is very pleasant, with good facilities and a spectacular view of the Serra da Estrela. There is also a caravan for hire for 4 people. There is a small swimming pool, but for great swimming be sure to take a dip in the river 3 km away. Campers may use the refrigerator.

The mountainous area has many old water mills, olive mills, churches and chapels that are worth a visit. Serra da Estrela national park starts at Unhais da Serra, which also boasts healing springs. The lively village of Paúl is also interesting. It has a renovated water mill and a museum at the Casa Típica, where you can listen to traditional songs accompanied by an *adufe* drum. The area has various villages dating back to Roman times. Festivals are held here in July and August.

How to get there

By car: Paúl is 20 km SW of Covilhã and 9 km S of Unhais da Serra. From Paúl, take main road to Tortosendo. Just past Taliscas, turn left. Farm is 300 m down this road. From Covilhã, take the main road to Tortosendo. At the intersection where the road continues to Unhais da Serra, turn left towards Paúl. Taliscas is 3 km down this road. Once in the village, after 50 m turn right on to a dirt road. Farm is 300 m further down. **Public transport:** Train to Tortosendo or Covilhã, thenbus to Paúl. From there, taxi.

28. Quinta da Cava

Francisco Pereira Matias
Taliscas, Paúl
6215 Unhais da Serra
Tel: 275-96 13 43
Open Jun-Oct Language: PT
Camping (tents only): see page 115

Farm and surroundings

This small traditional farm on the southern flanks of the Serra da Estrela is typical of the mixed farms of this region. In its fertile fields there are fruit and olive trees, vineyards and a vegetable plot. The livestock consists of cows and chickens. The friendly, elderly farmer and his wife are happy to involve their guests in their daily chores and provide them with many of the delicacies that nature has to offer. The sanitary facilities for campers are excellent.

The farm property is bordered by the small Paúl river, featuring waterfalls, old water mills, olive mills and Roman bridges. The river makes for excellent swimming and rowing (2 rowing boats are available). The mountainous, forested surroundings are good for walking. There is a renovated water mill and a museum in the Casa Típica, where traditional songs are performed. Erada, Casegas and Couto Minheiro are villages that date back to Roman times. They host several festivals in the months of July and August.

How to get there

By car: Paúl is 20 km SW of Covilhã and 9 km S of Unhais da Serra. From Paúl, take main road to Tortosendo. Just past road to Erada, cross river. After 2.5 km, you reach Taliscas. Just before sign indicating end of village, turn right on to unpaved road. Continue 0.5 km to farm. From Covilhã, take main road to Tortosendo. At intersection where road continues to Unhais de Serra, turn left to Paúl. After 3 km, you reach Taliscas. Just past sign indicating village limits, turn left on to unpaved road. Continue 0.5 km to farm. **Public transport:** Train to Tortosendo or Covilhã, then bus to Paúl. Here, taxi.

29. Quinta Vale Parais

Josef Schreiber & Anja Flier
Apartado 1, 6060 Idanha-a-Nova
Tel: 277-91 42 04
Open all year Language: PT, GB, D, F
€ pn Camping: tent 2.50-3; adult 2.50; child 1.50; Self-catering: 35; platform 5; dinner 8

Farm and surroundings

This very special farm is situated on a peninsula in the Barragem da Idanha, a fresh water reservoir with sandy beaches. Sheep and horses graze on the 50 ha property, which is planted with olive, oak and fruit trees and an organic vegetable garden. The farm is set in an open, green landscape.

The friendly farmers rent out a self-catering apartment with an open fireplace and completely furnished trailer home with a view of the lake (both max. 5 people). In the hot season, guests may sleep on platforms built in the trees. The lake shore offers plenty of space for campers to pitch their tents. Solar panels gen-

erate the electricity.

Experienced riders can join organized excursions lasting several days. On these trips, participants sleep in tents. Short introductory courses on horse husbandry and riding lessons are also available (information at the farm or ECEAT Portugal). Boats and bicycles are for hire.

Otters, black storks and vultures live near the dam. Beautiful wild flowers grow on the banks of the river Ponsul. Upstream you will find Idanha-a-Velha, a village that dates back to Roman times. Monsanto, known as 'the most Portuguese village in all of Portugal' is also worth seeing.

How to get there

By car: Idanha-a-Nova is 45 km NE of Castelo Branco. In Idanha-a-Nova follow signs to Idanha's campsite. Here follow signs and blue dots on stones and trees. Leave car at 'P'-sign and cross a creek (dry in summer) and in 10 minutes you'll arrive at the farm. If it's raining and in winter-time you will be picked up at the creek in a canoe, if you call in advance. **Public transport:** Train to Castelo Branco, then bus to Idanha-a-Nova. Paid pick-up service from Idanha-a-Nova or Castelo Branco (only at weekends).

30. Camping PortUgo

Wim & Corry Stok
Bardadeiro, Pampilhal, 6100-297
Cernache do Bonjardim.
Tel: 274-80 22 98
bert.camstra@wolmail.nl
Open Apr-Sep Language: NL, GB, D
€ pn Camping: tent 2-2.50; caravan 4; adult 2.10; child 1.30; caravan to rent 40; breakfast 4; dinner 9

Campsite and surroundings

PortUgo is a small and very quiet campsite, far from the hustle and bustle of city life. It is situated in the rural heart of Portugal, an area where there is both unspoilt nature and cultivated land.

Guests may choose to stay in a comfortable caravan (for 4 people) or in one of several fully furnished hire tents. Breakfast, lunch and dinner are served in a small open-air restaurant. There is no traffic, so children can play safely. Pets are also welcome. There is a mini shop, which sells basic necessities. Walking itineraries and maps are available, and bicycles, boats and canoes for rent.

PortUgo is an excellent departure point for walking, cycling and canoeing on the Zêzere lakes and for day trips to historical towns such as Tomar, Castelo Branco, Abrantes, Fátima and Batalha.

How to get there

By car: Pampilhal is 3 km N of Cernache do Bonjardim. From Cernache do Bonjardim, go N to Figueiró dos Vinhos (N237). After 2 km, turn left towards Pampilhal. Continue straight at the intersection with the traffic mirrors and turn right after 200 m. Follow the signs to 'PortUgo'. **Public transport:** Train to Tomar, then a bus to Cernache do Bonjardim. From there, take a taxi or call in advance for paid pick-up service.

Estremadura, Ribatejo and Alentejo

Monsaraz

Estremadura

No other part of Portugal can be compared to the fertile and varied landscape of the two former provinces of Estremadura and Ribatejo. This region is characterized by a great cultural and economic diversity. In Estremadura, the highlands are used for traditional agriculture and herding sheep and goats. The hilly land in the west is used to grow citrus fruit, olives and grapes. In the south, the densely wooded Serra de Sintra's wet ocean climate and (sub)tropical trees create a splendid, almost decorative landscape. This area, with its warm springs in Caldas da Rainha, Estoril and elsewhere, is one of Portugal's main tourist attractions. In the Tejo, Vouga and Mondego estuaries, you will find old fishing harbours and intensive rice cultivation. There are three nature reserves: the Tejo and Sado estuaries and the Parque Natural da Serra da Arrábida, which has splendid walking paths through subtropical vegetation. Places to visit include Alcobaça, famous for its beautiful Cistercian abbey; Batalha with its richly ornamented cloister; Peniche, a fishing village perched on a rocky peninsula and Nazaré, both on the Costa da Prata); and Sintra with its Moorish castle which was converted into a beautiful palace, surrounded by flower gardens and parks. And of course, there is Caldas da Rainha, whose hot sulphur springs have made the town one of Portugal's most famous spa resorts. Obidos is a fortified town where time seems to have stood still. And last but not least, there is nostalgic Lisbon. As the saying goes: *quem não viu Lisboa não viu Portugal* (whoever has not seen Lisbon, has not seen Portugal).

Ribatejo

This region is named after the banks of the river Tagus. Appropriately so, for it is this river and its unpredictable course which determine the landscape of Ribatejo, Portugal's most fertile region. South of the Tejo are expansive pastures where bulls, cows and horses graze. The towns of Santarém and Vila Franca de Xira are known for the parades of local *campinos*, the cowboys of the Ribatejo. The region's north-east is dominated by intensive farming of vegetables, grain and fruit, including olives and wine grapes. The central and southern parts of the region abound with wheat fields, olive trees and cork oaks. Here, the landscape begins to resemble Alentejo.

Towns worth visiting include the pilgrimage centre of Fátima (Portugal's answer to Lourdes) and beautiful Tomar. Here, the mediaeval Knights Templars and Knights of Christ built the Castelo dos Templários and the Convento de Cristo respectively.

Alentejo

This southern province is a seemingly endless plateau south-east of the river Tejo. Alentejo is divided into a northern half (Alto-Alentejo) and a southern half (Baixo-Alentejo). Both are sparsely populated; the terrain is flat and rough. Great expanses of grain are interspersed with plains of cork oaks and olive trees, cut through by many densely wooded valleys. The area also has several mountain ranges: Serra de Cercal, Serra de Grândola and Serra de São Mamede. Alentejo's west coast has unusual rock formations and sandy beaches. In 1995, this coastal area was declared a nature reserve, called Sudoeste Alentejano.

Alentejo has a long history of human habitation. Several traces of prehistoric settlement have been discovered in Alto-Alentejo. There are megalithic tombs, dolmen, menhirs, and a cave with prehistoric drawings (Gruta do Escoural). These finds are concentrated near Montemor-o-Novo, Évora and Monsaraz. There are also numerous remains from the periods of Roman and Moorish rule. In Alentejo, feudal conditions survived long after the Middle Ages. During the 1974 Carnation Revolution, landless migrant workers spontaneously occupied large properties and many co-operatives were founded. However, most land expropriations were reversed in later years and many co-ops became defunct.

In Roman times, Alentejo was the granary of the Empire. Agriculture, both crop farming and cattle breeding, is still the mainstay of the region. Intensive farming has badly eroded some areas, leaving only a prairie-like landscape. Alentejo still produces wheat, rye and corn, but visitors will also come across herds of sheep or goats, their shepherd assisted by his dogs. With a little luck, you will also see a herd of black Alentejan pigs foraging for food.

Alentejo is dominated by big farms, known as *latifundia* or *herdades*, which are owned either by large landowners or farmers' co-ops. The accommodation in Alentejo can be booked year round. However, the summers are extremely hot and dry. The best times of year to visit Alentejo are around Christmas, in the spring and in late summer. Winters are very mild (15°C on average) and usually rainy, but broken up by beautiful, sunny days.

Almourol castle, Constância

31. Campismo O Tamanco

Irene van Hoek & Hans de Jong
Casas Brancas 11, 3100-231 Louriçal
Tel/fax: 236-95 25 51
campismo.o.tamanco@mail.telepac.pt
www.campismo.o.tamanco.com
Open all year Language: NL, GB, F
€ *pn Camping: tent 2; caravan 2.75; camper van 4.25; adult 2.50; child 1.25; Self-catering pw: 128(ls)-256(hs); dinner 7.25; B!*

How to get there

By car: Louriçal is situated between Figueira da Foz and Pombal. Coming from Pombal, you will arrive at a roundabout in Louriçal. Turn left. Follow the road for 5 km until you reach the campsite (on your left). Coming from Figueira da Foz, take the N109 in the direction of Leiria. Take the 2nd exit to Louriçal. After 1.5 km, you will find the campsite on your right-hand side.
Public transport: Train to Louriçal (3 km from the campsite). Or take a bus Leiria - Figueira da Foz, which stops 800 m from the campsite.

32. Casa da Maria Pequena & Casa Trindade

Gerard Driessen
Madroeira, Bêco
2240-222 Ferreira do Zêzere
Tel: 236-63 17 67 / (024-322 32 26 in NL)
lydiapvw@clix.pt
Open all year Language: NL, D, GB
Self-catering pw: €178-€223(ls)
€223-€255(hs)

Campsite and surroundings

The 'O Tamanco' campsite is situated on 1.5 ha, partly covered with fig, mimosa and various fruit trees as well as bougainvilleas. There is plenty of space for tents and caravans, and the campsite has a swimming pool, playground and good sanitary facilities. A small restaurant called 'A Cantina' serves delicious meals with a choice of vegetarian, meat or fish dishes. You can also purchase local products including freshly picked produce. This is an environmentally friendly campsite, where waste is separated, organic matter is turned into compost and biodegradable products are used. The comfortable common room has a collection of books and magazines and a television. Outside there are several shady terraces. The owners organize workshops in sculpture, painting and Reiki. For more active guests, maps of hiking and cycling routes are available.

Houses and surroundings

Casa da Maria Pequena is situated on 1 ha of terraced land with a stream running along one side. In the traditional wood-burning oven the Dutch owner has baked a lot of bread and cakes over the years. The property is dotted with vines, olive trees and fruit trees: apple, pear, cherry, orange, lemon, peach and fig, irrigated with water coming from 2 large storage tanks (restored). Drinking water is obtained from a well. The house has enough room for 4 guests.

Casa Tridade is situated 8 km further on, in the tiny village of Alfqeidão. From the veranda you can enjoy the sunset and the view over the fruit trees and vines. Both houses have been re-built using natural materials and are heated by woodburning stoves.

For great swimming and canoe/rowing boat rentals, go to the mighty Zêzere river (5 km away). There are many small villages nearby, each with its own folklore and summer festival. A weekly market is held in Cabaços. Tomar, a lovely historic city, is 30 km away.

How to get there
By car: Bêco is 25 km NE of Tomar. In Cabaços, between Coimbra and Tomar on the N110, take the exit towards Ferreira do Zêzere/Portela do Braz. After 2 km, turn left to Fonte Seca and Bêco. After 3 km (at a crossroads with big green container to your left) turn left again and after 100 m take another left turn. Drive another 500 m over unpaved road to the lodg-ing. It is the last house in the hamlet. **Public transport:** Train to Tomar (at 25 km) and then the bus to Ferreira do Zêzere and Cabaços. From there take a taxi to the village of Madroeira.

33. Camping Redondo

Hans Frommé
Poço Redondo, 2300 Tomar
Tel/fax: 249-37 64 21
hansfromme@hotmail.com
http://home.wanadoo.nl/edvols
Open Oct-Sep; camping Mar 15th-Oct
Language: NL, PT, D, F, GB
€ pn Camping: tent 2.25-3.25; caravan 3.25; camper van 3.75 adult 2.80;
Self-catering: 37-55(ls) 48-65(hs) B!

Campsite and surroundings
This small rural campsite with its citrus trees is located in the village of Poço Redondo (Round Well), 10 km from historic Tomar and 5 km from the beautiful Zêzere lakes. A hand-paint-ed script on the gable of the house, which dates from 1922, reads: *Bem vindo seja quem vier por bem!* (Welcome to those who come for the good). The rolling hills, with their olive, or-ange, lemon and fig trees, offer a pretty view of the surrounding vineyards. The original ser-vants' quarters have been converted into a self-

catering apartment for 4, and there are 3 log cabins and a caravan (also available in winter). There is a swimming pool, a huge draughts board, a playground and a grassy field for vol-leyball and badminton, a *jeu de boules* and table tennis. Mountain bikes and canoes are for hire.

On arrival guests receive a folder with infor-mation on walking and biking itineraries, cul-tural sites and beaches along the Zêzere lakes. The convent in Tomar is well worth a visit.

How to get there
By car: Poço Redondo is 10 km from Tomar. On the N110, coming from Lisboa or Coimbra, take just before Tomar the IC3. On this ring road, exit towards Albufeira do Castelo de Bode - Tomar. From here green signs with 'Camping Redondo' & red hearts will show the way (7.2 km to go). **Public transport:** Train to Tomar. Then bus to Poço Redondo (3x daily, stops at 300 m). Paid pick-up service from Tomar.

34. Parque de Campismo Rural da Silveira

António Viegas & Nicole Laurens
Capuchos, 2460 Évora de Alcobaça
Tel: 262-50 95 73
silveira.capuchos@clix.pt
Open all year Language: PT, GB, F
€ pn Camping: tent 2.50-3.50; caravan 3.50; camper van 3.50-4.50; adult 2.50; child 1.25; dinner 5

Farm and surroundings
Silveira is a small campsite on an 8 ha farm with orchards, vineyards, olive trees, cork oaks and pine trees. Owners António and Nicole take

pride in maintaining the farm's peaceful, natural atmosphere.

The campsite has 18 pitches, separated by natural hedges to ensure privacy. Campers have access to a common room, a barbecue area and a washing machine. The showers and bathrooms are clean and well-maintained. The campsite is very suitable for children, with plenty of space to play, run and ride bicycles in safety. Advance booking is necessary from Oct. 1 to May 31.

Situated between Alcobaça, the Costa da Prata (Silver Coast) and the Serra de Aire e Candeeiros, the campsite is ideally located for those wishing to explore Portuguese culture, the coast or the mountains. There are various marked footpaths in the Serra, and you can also take an excursion on the back of a donkey.

How to get there

By car: Évora de Alcobaça is 7 km from Valado. From Alcobaça, follow signs to Évora de Alcobaça (N86). After 3 km, you will see a gateway / archway flanked by white walls on the left-hand side. This is the entrance to the campsite. **Public transport:** Train to Valado, then a bus to Alcobaça. Then take a taxi to the campsite.

35. A Colina Atlântica, Quinta das Maças

Ineke van der Wiele & Ton Kooij
Travessa dos Melquites 3, Barrantes, 2500
Caldas da Rainha
Tel/fax: 262-87 73 12 Mob: 936 72 49 58
*Open Mar-Nov Language: PT, NL, GB, D, F
Lodging pw in €: 1p B&B 175; 2p 225-280;
dinner 10*

Houses and surroundings

For Dutch couple Ineke and Ton it was a labour of love to create a place where guests can get back in touch with themselves and others. A Colina Atlântica is beautifully situated in the hills, just 8 km from the Atlantic Ocean. The house is surrounded by a spacious garden full of terraces, and with seats and hammocks. The apple orchard (1 ha) stretches up to the hilltop.

The main building houses a meditation loft, kitchen, bar and dining room. Guests can rent a double or triple room with private bathroom. There are also 3 caravans and a single room, which share a shower and toilet. Prices include breakfast and a morning meditation session.

Good dinners and barbecues are prepared on the patio. Notify your hosts in advance if you wish to join them for these meals. You can also join excursions to the monasteries of Alcobaça and Batalha, the rocky coastline of Peniche, various local markets and unspoilt beaches.

Guests may also take courses in Reiki and Yi Qing. Four times per year, at the turn of the seasons, Ineke and Ton organize a 'Meditation Intensive'. Groups and trainers wishing to teach their own groups are also welcome.

How to get there

By car: Barrantes is 6 km NE of Caldas da Rainha. From Lisbon, follow A1 Norte. After some km, exit for Ponte Vasco da Gama and keep left: 'Outras Direcções' (toll road). After Torres Vedras, the A8 changes into IC1 towards Caldas da Raihna. Exit for Tornada. At traffic lights in Tornada go right to Barrantes. At end of Barrantes keep left at fork and go right after 150 m. Farm is 50 m down path. **Public transport:** Train to Caldas da Rainha, then taxi or pick-up service (free of charge).

36. Casal do Pomar

Fredy Seitz
Bouro - Salir do Porto
2500 Caldas da Rainha
Tel: 262-88 13 59 Mob: 966-90 37 77
Open May-Oct Language: PT, GB, D, F
Self-catering pw: €215-€265(ls) €375-
€525(hs) B!

Houses and surroundings

Situated in a bird sanctuary, the village of Bouro is home to 3 families who live in close harmony with nature. Casal do Pomar consists of 4 fully furnished houses, each with a distinct architectural style. There are beautiful patios, herbaceous borders and secluded retreats in the organic orchards. There is plenty to do in this area for both children and adults: day trips, water sports, tennis, kite flying and riding, etc. Bicycles are available for hire. Beaches, lagoons and natural bays abound on the Costa da Prata. Further inland lies 'Portugal's green garden', where most of the country's fruits and vegetables are grown. The region is also famous for its ceramics and pottery.

In the town of Caldas da Rainha (The Queen's Baths), you can relax in thermal springs and visit museums to discover traditional pottery, sculpture and paintings. From Caldas da Rainha, it is only a 10 minute drive to Óbidos, town of castles. Those with an interest in history can visit the monasteries of Alcobaça, Batalha, Fátima and Tomar, which are an hour's drive from the lodgings. Near Fátima and Ourem, there are caves with stalactites, stalagmites and dinosaur footprints.

How to get there

By car: The hamlet of Bouro is 5 km NW of Cal- das da Rainha and 4 km from the Atlantic. In Caldas da Rainha, head for Alcobaça. After 3 km, turn left towards Salir do Porto. Drive through the village of Chão da Parada, then turn left at the sign to Casal do Pomar. Bouro/Casal do Pomar is 500 m down the road. **Public transport:** Train to Bouro, then walk the last 500 m. By bus from Lisbon (2x daily) or from Caldas da Rainha (4x daily) to Salir do Porto. From there it is another 2 km. Taxi from Caldas da Rainha. Pick-up service from the airport.

37. Moinho de Vento

Clotilde Veiga
R. Francisco Sá Carneiro,
Penedos de Alenquer, 2580 Alenquer
Tel: 214-74 02 68 Mob: 962-81 70 11
Open all year Language: PT, GB, F
Self-catering pn: €35; extra bed €5; B!

Accommodation and surroundings

Moinho do Vento ('Windmill'), is situated in the Estremadura, looking out over hilly vineyards dotted with villages and fields as far as the eye can see. The landscape is stunningly beautiful, especially at sunset when the light plays on the buildings. In the past, the mill was used to grind grain; today it is a renovated, rustic guest house, but inside you will still find all of the original gearing. Because of the mill's conical structure, its rooms all have different dimensions and plenty of character. It has a double bedroom and a living room with 2 sofa beds, a bathroom and a small kitchen. Guests can cook and dine in the small yard. The Montejunto mountains, the Berlengas islands and the towns of Óbidos and Caldas de Rainha are all well worth a visit.

How to get there

By car: Penedos de Alenquer is 18 km NW of Alenquer and 50 km NE of Lisbon. On the A1 Porto - Lisbon, exit at Carregado (exit 4) and to Alenquer. In Alenquer, take the road to Torres Vedras. After a few km, at the fork in the road, turn right. Pass through Olhavo and stay on this road until a sign for Labrugeira, Penedos de Alenquer and Abrigada. Here, turn right. After passing Labrugeira, continue towards Penedos de Alenquer. You will find Moinho do Vento on the left, just after the sign reading 'Penedos de Alenquer'. **Public transport:** Train to Torres Vedras (at 20 km). Bus to Penedos de Alenquer (stops at 1 km from mill). Taxi from Penedos de Alenquer.

grow vegetables, medicinal herbs and fruit. They use their home-grown organic produce in the vegetarian meals they serve.

There is a double room for hire, and a large range of organic and home-made products for sale at the farm. You can attend various courses, such as vegetarian cooking, basket weaving, organic farming and yoga. Lígia or António are happy to accompany you on walks, cycling tours or trips by car.

How to get there

By car: Bucelas is 25 km N of Lisbon. From Bucelas, head W to Malveira (N116). After 800 m, you see a tall, white building on your right. Across the road, on the left, a pillar reads 'Quinta das Lágrimas'. Turn left on to the unpaved road. Continue on this road: do not take any roads off to the left or right. Pass first residence on the right. Next house on your left is Quinta da Picota. **Public transport:** Train to Bucelas, then a Lisboa-bound bus. Get off at first stop (at a tall, white building). Or taxi from Bucelas. Pick-up service from Bucelas or Lisboa.

38. Quinta da Picota

José Joaquim & Lígia Maria
Casal da Torre de Baixo
2670 Bucelas
Tel: 219-69 36 15 Mob: 966-95 94 63
*Open all year Language: PT, GB, F, E
Lodging € pn: 2p 25; extra bed 7.50;
breakfast 2.50; dinner 7.50*

Farm and surroundings

Although Quinta da Picota is near Lisbon, it looks out over pristine hills abundant with wildlife and plants. Owners José and Lígia

39. Quinta do Pomarinho

Georg Gautier
Apartado 2, 7320 Castelo de Vide
Tel: 245-90 12 02 / 90 52 25
pomarinho@clix.pt
http://pomarinho.planetaclix.pt
*Open all year Language: PT, GB, D, F
€ pn Camping: tent 3.50; caravan 4.50;
adult 3.50; child 1.75; hut 5; Lodging: 2p 20*

Farm and surroundings

Set in a remote and hilly area, this organic farm has donkeys, cows, sheep, chicken, bees and cats. The 25 ha property consists of grassy meadows, olive trees and cork oaks. Its friendly owner is retired but still active in agriculture and social development.

The lodgings are tended by two young Dutch women. Guests can stay in the old (renovated) house, in a new bungalow, or in one of the simple African huts on the premises. Guests who stay in a hut eat their meals in a pavilion. On the property you will also find a vegetable garden, a natural pool and a small swimming pool. Children are allowed to ride the donkeys.

The farm is near the beautiful Serra de São Mamede nature reserve. Castelo de Vide still bears the marks of the various civilizations it has seen. The 13th century castle reigns supreme. The historical centre of Mavão and the lively town of Portalegre are also worth a visit, as are the Pavoa and Meadas lakes.

How to get there

By car: Castelo de Vide is 13 km from Portalegre, near the Spanish border. Driving from Portalegre, there is a sign to Quina do Pamarinho on the right-hand side of the road. **Public transport:** Train to Castelo de Vide. From station taxi or call to be picked up (free).

city of Évora. The granite, brick and plaster farmhouse was built in the early 19th century. Owners Laura, João and their 3 children offer their guests a pleasant stay in a warm family atmosphere. The farm has a small vegetable plot and olive and fruit trees. The hosts teach macrobiotic cooking and organize Reiki sessions.

The Alentejo plains abound with villages and churches of cultural and historical interest. Évora, a world heritage city, has a variety of cultural events on offer almost the whole year round.

How to get there

By car: Canaviais is 4 km N of Évora. From Évora, head for Estremoz. Across from bar/restaurant Parreirinha on your right, turn left towards Canaviais. Continue on this road for approx. 4 km. The farm is the next to last property on your right before you enter the village. The entrance is a narrow gate between stone walls. **Public transport:** Train to Évora, then bus on Praça do Giraldo to Canaviais (get off across from Café Inácio). Or take a taxi from Évora.

40. Quinta das Chaves

Laura Dinis & João da Silva
Canaviais, 7000-213 Évora
Tel: 266-76 15 19
Open all year Language: PT, GB
B&B pn: 2p €40; extra bed €10; dinner €10

Farm and surroundings

Quinta das Chaves is one of several old-fashioned farms in the countryside surrounding the

41. Quinta Bica da Matinha

Hans & Kate Groenendijk
Bica da Matinha, 7555 Cercal do Alentejo
Tel: 269-90 48 81
Open all year Language: NL, PT, GB, D
Self-catering: €40 pn

Farm and surroundings

Bica da Matinha farm is situated in a small and isolated valley in the Serra do Cercal hills. The property has 35 ha of cork oaks and pines overlooking the sea. The Dutch owners - Hans, a homeopathic doctor, his wife Kate and their children - have renovated their Alentejo-style house, taking great pains to preserve its original style. The family has cats, dogs, chickens and pigeons.

They have 2 small apartments for rent, each large enough for 2 adults and 2 children. Guests share a small kitchen located outside the apartments.

There are horses for hire and guided riding tours can be arranged. Kate teaches watercolour painting.

Because of its location on the Atlantic coast, this farm is ideal for people who love deep sea fishing. It is also recommended that you walk in the dunes of the Costa Vicentina nature reserve and visit the numerous fishing villages along the jagged cliffs of the Alentejo coastline.

How to get there

By car: Cercal do Alentejo is 30 km from Avalade. On the road from Cercal do Alentejo to Milfontes (N390), you will come across a white factory building (after 1 km on your left). Across the street and on your right is an unpaved road. Take this road, following the telephone poles that run along it. At the crossroads, keep going straight. At the next three-pronged fork, take the left road. This will bring you straight to the farm. **Public transport:** By train to Avalade, then by bus to Cercal. From there, take a taxi.

42. Quinta Cerca dos Sobreiros

Inka Killing & António Aires
Montes Altos - Santana de Cambas
Mina de São Domingos, 7750 Mértola
Tel: 286-64 73 52
Open all year Language: GB, PT, D
Camping: see page 115

Farm and surroundings

Quinta Cerca dos Sobreiros is a characteristic Alejento farm: the low white buildings are set in a terrain dominated by cork oaks and olive trees. The farm is situated in the little valley of a brook that dries to a trickle in the hot summer months. The area borders directly on Spain, and is one of the most deserted but beautiful regions of Alejento. Until the 1960s,

the major source of income in this region was mining. When the mines closed, most people left the area. At present, there is a growing interest in restoring the dilapidated buildings and organizing activities that will keep the local population from migrating. Inka, António and their daughter are not only revitalizing their farm but are also involved in various other activities. António deals in Oriental arts and crafts and Inka is a great vegetarian cook who organizes summer camps for children. There is one canoe for hire.

The campsite has room for 10 tents and 3 caravans. The surrounding area is ideal for riding, cycling and swimming (in the river and lakes). The region has various interesting sights to see, such as Guadiana nature park, an archaeological park, a Moorish castle and an archaeological museum.

How to get there

By car: Mina de S. Domingos is 70 km SE of Beja and 17 km E of Mértola. In Mina de S. Domingos across from the church, take the road to Montes Altos. You will pass an old mine crater: 'the black hole'. Approx. 2 km out of Mina de S. Domingos and 100 m before you enter the village of Montes Altos, turn right on to a dirt road going slightly downhill. The farm is 300 m down this road. **Public transport:** By train to Beja, then by bus to Mina de S. Domingos.

Ria Formosa

The Algarve

Like many southern Portuguese names, the word 'Algarve' was inherited from the Moors: *al garbh* (the West) refers to the region's location in the mediaeval Islamic world. The Algarve has seen many peoples come and go. The favourable climate and fertile soil attracted Phoenicians, Greeks, Celts, Carthaginians, Romans, Visigoths and finally the Moors, who stayed for a full 500 years. The Algarve was the last region to become part of modern Portugal.

Shielded by the Monchique and Caldeirão mountains, the Algarve has a subtropical climate and rocky beaches which have made it Portugal's number one tourist attraction. The region is relatively quiet in spring and autumn. The rocky coastline in the west has been developed more than the sandy eastern shore. Along the latter, you can still find romantic little fishing villages. In the relatively humid area inland, there are green orchards, vineyards and even rice paddies and sugarcane fields. The climate and soil provide excellent conditions for growing almond, fig and olive trees. On higher slopes there are forests of eucalyptus, pine, cork oak and strawberry trees (*arbutus unedo*), whose fruit is used to produce *aguardente*, a famous regional brandy.

Hidden in the vast landscape of wooded hills and mountains are some quite beautiful little villages. Unfortunately, many mature forests have made way for eucalyptus plantations. Travellers on the network of dirt roads will encounter many donkeys, traditional wells using animal power, and isolated farms.

Most of the farmers running the accommodation included in this guide are not Portuguese,

but foreigners. This speaks volumes about the rise of organic farming in Portugal. The Algarve is home to a group of active farmers who have joined forces and pooled their resources, sharing technical advice, giving demonstrations, organizing local distribution and the national and international marketing of their products. They are one of the driving forces behind Portugal's organic agriculture. Perhaps they will eventually inspire the original Portuguese farmers as well. Of the small handful of Portuguese farmers who have survived until today, only a few are involved in organic farming. ECEAT Portugal tries to support these farmers and help them develop the concept of eco-tourism.

43. Olhos Negros

How to get there

By car: In Monchique, take the N266 to Sabóia and Lisbon. After 5 km, you will see Restaurante Típico 'O Poço da Serra' on your right; 50 m further, turn left on to an unpaved road. Follow the signs for 7 km until you reach the farm. Caution: the road is sometimes barely passable.
Public transport: Train to Portimão, then bus (11x daily) to Monchique. Pick-up service from the restaurant (€3) or from Monchique (€6).

Stichting Desh
Apartado 8, 8550-909 Monchique
Tel: 282-95 52 07 (020-468 74 03 in NL)
Mob: 933-46 86 93 (065-153 22 18 in NL)
Fax: 282-95 52 09
desh@ip.pt www.algarvenet.pt/desh
Open all year Language: NL, PT, GB, D
€ pn Camping: adult 6.50; child 3.25;
Lodging: 2p 20; Bunk house 160

44. Quinta Vinha Velha

Hubert Muller & Margit Kegel
Barão de São João, 8600 Lagos
Tel: 282-68 72 86
Open all year Language: PT, D, GB
€ pn Lodging: 2p 20-25; Self-catering:
27.50(ls)-40(hs) B!

Farm and surroundings

This farm, once a famous *medronho* distillery, was converted into a group lodging for youngsters. However, anyone interested in peace and quiet may rent the building, now owned by the Dutch Desh foundation, which apart from offering a nice holiday on the spot, organizes gay vacations and bus tours. The farm is set in a very sparsely populated, heavily wooded area.

There is a common room with an open fireplace, a bar and a dining room. There are 2 dormitories (for 12 and 20 guests) and a kitchen. The patio has a barbecue area and a Portuguese oven. The terraced fields near the farm offer plenty of space to pitch your tent. Electricity is generated by solar panels and drinking water comes from a well on the property. Nudity is allowed. The Desh foundation organizes a range of group activities, excursions and trips. The area is very suitable to explore on foot. The plant life is very diverse: cork oaks, *medronheiros* and eucalyptus. Local animal life includes wild boar, eagle owls, eagles and partridges.

One of the reservoirs nearby is suitable for swimming and paddling in rubber boats. Advance booking is necessary. Gays are welcome.

Farm and surroundings

This is a mixed organic farm in the middle of a wide and mountainous landscape that stretches all the way to the west coast. There are cows, sheep, horses, a donkey, cats and dogs. The milk is processed into cheese and yoghurt and a large vegetable garden supplies the necessary greens. The German owners - Hubert, Margit and their children - strive to live as naturally as possible. They have renovated 2 old farmhouses and built 2 new wooden houses using environmentally friendly methods and materials (such as sheep's wool as thermal insulation).

They rent out various double bedrooms as well as 2 self-catering apartments for 4. One of these is on the shore of a lake, 500 m away from the farmhouses. Reservations are necessary.

The donkey and horses are for hire. The surrounding area is ideal for long walks through the fields and woods. The beach is 10 km away. The traditional village of Barão de São João is hospitable to visitors.

How to get there

By car: Barão de São João is 15 km NW of Lagos. In Lagos, head for Aljezur/Lisboa. Just past Portelas, turn left towards Barão de São João. Upon entering the village, take an immediate left towards Sagres. After 100 m, at a water tap, turn right. The road goes uphill into the protected woods (Mata Nacional). At the 3rd intersection (Vinha Velha), turn left. After 100 m, you will leave the Mata Nacional. The road leading downhill off to your right will take you to the farm. **Public transport:** Train to Lagos, then a bus to Barão de São João. Paid pick-up service from Barão de São João.

---

45. Monte Rosa

Sandra Falkena
Barão de São João, 8600-016 Lagos
Tel: 282-68 81 78 / 68 70 02
Fax: 282-68 70 15 monte.rosa@clix.pt
www.monterosaportugal.com
Open all year Language: NL, PT, F, D
€ pn Camping: caravan 6; adult 5; tent to rent 4; Lodging: 1p 35-45; 2p 45-50; 4p 60; Self-catering: 75-100; extra bed 11; breakfast 3; dinner 10

Monte Rosa and surroundings

Monte Rosa's small guest lodgings and campsite are surrounded by natural beauty in the SW corner of the Algarve. It is run by Sandra, a Dutch woman who has lived there with her

daughter Arianne since 1993. There are rabbits, chickens and cats living on the premises. The 3.6 ha property has almond, fig and olive trees. There is plenty of space to pitch your tent or park your caravan. There are 2 carefully renovated buildings and a small campsite with good facilities and an outdoor kitchen. There are single, double and quadruple bedrooms for rent, as well as a self-catering apartments for 4 and 6. There are also tents, mattresses, duvets, bed linens and towels for hire. The breakfast restaurant with patio also serves dinner several days a week. The complex is surrounded by a colourful garden with several little pathways, terraces and secluded sitting areas. You can relax on a sundeck and swim in a salt water pool, with panoramic views across the Portuguese countryside.

Paid guided walking tours of the area can be arranged. Monte Rosa has information on 20 walking routes, and arranges transport to and from the starting points and a packed lunch. Day trips, riding and bicycle hire and massages are also available. There is a library and a playground for children. Monte Rosa can also be hired to groups for workshops, meetings and courses.

How to get there

By car: Barão de São João is 10 km NW of Lagos. In Lagos, head for Aljezur/Lisbon. Just past Portelas, turn left towards Barão de São João. After 6 km, there is a sign to 'Monte Rosa' on the left-hand side. **Public transport:** By train to Lagos, the by bus to Barão de São João. Or take a taxi from Lagos. Paid pick-up service from Lagos (€5) or Faro (€40).

---

46. Casa Stakwaas

Huub Zonneveld & Ineke Beets
Sítio do Mendronhal, Barão de São João, 8600 Lagos
Tel: 282-68 72 10
Open all year Language: NL, PT, GB, D, F
Self-catering pw: €175-€375

Apartments and surroundings

Owners Huub and Ineke built Casa Stakwaas by hand in 1985. Set in a quiet, wooded and rural part of the Algarve, the house is 12 km from the S coast and 15 km from the W coast. Huub is an electronics repairman, and also builds irrigation and solar power systems. Ineke buys and sells used Portuguese furniture and home decoration items. She and her sister own a shop in Barão de São João, 1.5 km from Casa Stakwaas. Barão de São João is a typical Algarve village where you can get acquainted with the Portuguese way of life.

Casa Stakwaas is a semi-detached house. Huub and Ineke live in one half; the other is rented out as a self-catering apartment for 4. There is also a wooden cabin large enough for 2 guests. The lodgings are a good place to begin walking, cycling and riding tours. The closest beach is Praia da Luz, 6 km away.

How to get there
By car: Barão de São João is 15 km NW of Lagos. **Public transport:** By train to Lagos, then by bus to Barão de São João. Get off at the bus stop 100 m before restaurant 'O Cangalho'. Take the 1st paved road to your left. Casa Stakwaas is 500 m down the road, on your right.

47. Quinta Aloé Vera

Karin Giesewetter
Vale de Meias, Barão de São Miguel
8600-013 Lagos. Tel: 282-69 58 82
Open May-Nov Language: PT, D, GB
Camping (tents only): see page 115;
€ pn Lodging: 1p 15; 2p 25; Self-catering: 35;
extra bed 10

Farm and surroundings
An Angolan man, a German woman and their 4 children are turning this beautiful property into a viable enterprise. The wooden house was built using environmentally friendly methods and materials; for example the wood was treated with aloe vera. The house sits on a hill, surrounded by cork oaks and beautiful herbaceous borders, vegetable plots, fruit trees and aloe plants. Karin uses the aloe vera for the medicinal and cosmetic products she makes. A donkey, chickens, dogs and cats all live on the farm.

The 16 pitch campsite is in a lush area near a small lake. Sanitary facilities are basic: there is a compost toilet and the bathrooms are simple and have cold water only. Part of the house is rented out as a self-catering apartment for 6, with its own living room and kitchen (bring your own kitchen gear). This place to stay is very well suited to those interested in peace and quiet, simplicity and natural healing. There is playground and a small lake in a valley full of birds and wild boar.

The farm is 6 km from the Atlantic Ocean. It is a great spot to begin exploring the coastal region or the Algarve hinterland.

How to get there
By car: In Lagos, take the N125 to Vila do Bispo. Just past Almadena, turn right. This road brings you to Barão de São Miguel. As you drive out of this village, take the first turning on the right. Follow the signs to the farm (1.6 km). Better still: call the farm and you will be escorted there. Public transport: Train to Lagos. From there, take a bus to Barão de São Miguel. Paid pick-up service from Faro.

48. Quinta do Coração

Nicolette & Jon Gomes
Carrasqueiro, Salir, 8100 Loulé
Tel: 289-48 99 59
info@algarveparadise.com
www.algarveparadise.com
Open all year Language: NL, GB, F, D
€ *pn B&B: 1p 35; B&B 2p 45; extra bed 5;*
Self-catering: 45; dinner 10; B!

Farm and surroundings

Quinta do Coração - Farm of the Heart - is a charming country guesthouse set in peaceful surroundings. This truly unspoilt part of the Algarve is a paradise for hikers, birdwatchers, nature lovers and all who seek to relax and revitalize their minds and bodies. Besides the great variety of plants and trees here, there are also many indigenous birds such as the bee-eater, Sardinian warbler, Bonelli's eagle and green woodpecker.

The traditional Portuguese guesthouse was rebuilt using natural river stones, cane and wood. The farm has a small vegetable garden, a chicken run and some fine terraces. The owners - a Dutch-Jamaican couple with a young son - serve wonderful breakfasts and dinners on request (vegetarian, meat and fish). Wood-burning heaters provide warmth in winter. Nearby attractions include natural springs, historic sites, and picturesque rural villages and markets featuring basket weaving and leatherwork. The beaches of the Algarve are only half an hour away. From there, one can take a boat out to the small islands just offshore. Quinta do Coração is also open to groups, family reunions and workshops.

How to get there

By car: On the A2 S, exit at Messines. Head for Alte (N124). After Alte, head for Salir/Barranco de Velho. Follow road until sign to Carrasqueiro/Cabeça de Vaca, just before the bridge. Turn left on to unpaved road. Follow signs for 1.8 km to the farm. **Public transport:** Train to Loulé (20 km from farm). Bus from Loulé (twice daily) stops 2.5 km from farm. Paid pick-up service from Salir, Loulé and Faro.

Vegetarian and macrobiotic restaurants in Portugal

Porto

Suribachi
R. do Bonfim 136, E 140
Porto
Tel: 22 56 28 37

The Beiras

Bionatura
Rua de Outubro 143
Viseu
Tel: 232 42 96 88

Sonatura
R. Clube dos Galitos, 6
Aveiro
Tel: 234 38 16 01

Casa Pombal
Rua das Flores, 18
Coimbra
Tel: 239 83 51 75

Ribatejo & Lisboa

Macromar
Trav. do Vasco 10
Tomar
Tel: 249 31 34 25

Celeiro Dietético
R. 1º de Dezembro, 47
Lisboa
Tel: 21 342 24 63

*Centro de Alimentação e
Saúde Natural*
Rua Mouzinho da
Silveira, 25
Lisboa
Tel: 21 315 08 98

Chacra
R. Augusto Machado, 5-1
dto
Lisboa
Tel: 21 848 04 61

A Colmeia
R. da Emenda, 110-2º
Lisboa
Tel: 21 347 05 00

Espiral
Praça Ilha do Faial, 14A
Lisboa
Tel: 21 315 38 72

Hare Krishna
R. Dona Estefania, 91
Lisboa
21 314 03 14

Instituto Kushi
Av. Barbosa du Bocage, 88
Lisboa
Tel: 21 77 68 54

Nori
Av. Miguel Bombardia,
nº 145 r/c
Lisboa
Tel: 21 31 51 10

O Sol
Calçada do Duque, 25
Lisboa
Tel: 21 36 56 96

*Soc. Portuguesa de
Naturalogia*
Rua do Alecrim, 38-3
Lisboa
Tel: 21 346 33 35

Terraço do Finisterra
R. do Salitre, 117
Lisboa
Tel: 21 352 20 38

Tibetanos
Rua Salitre
Lisboa
Tel: 21 314 20 38

Unimave
R. Mouzinho da Silveira, 25
Lisboa
Tel: 21 355 73 62

Viragem
Travessa das Parreiras, 76
Lisboa
Tel: 21 53 33 60

Yin Yang
R. dos Correiros, 14-1
Lisboa
Tel: 21 342 65 51

Bem Me Quer
Av. do Cristo Rei, 23A
Almada
Tel: 21 274 40 54

Amanhecer
R. dos Combatentes da
Grande Guerra, 40
Barreiro
Tel: 21 207 86 63

O Bago de Arroz
R. António Joaquim Granjo
Setúbal

Algarve

Casa Primavera
Cruz Da Assumada
8100 Louie

Ilha da Madeira

Bio-Logos
Rua Nova de S. Pedro, 34
Funchal
Tel: 291 368 68

Useful addresses in Portugal

Tourist offices for Portugal

The Netherlands
Portugese ambassade
Afdeling Handel & Toerisme
Haagsche Bluf 63
2511 AR Den Haag
The Netherlands
Tel: +31 (0)70 326 25 25 / (0)70 326 43 71
Fax: +31 (0)70 328 00 25
icep.haia@icep.pt
www.portugalinsite.pt

United Kingdom
Portuguese Tourist Board
22-25 A Sackville Street
London W1X 1DE
United Kingdom
Tel: +44 (0)207 494 14 41
Fax: +44 (0)207 494 18 68
iceplondt@aol.com
www.portugalinsite.pt

Organic Agriculture

Certifying organization
Socert
Certificação Ecológica, Lda
R. Alexandre Herculano, 68-1° Esq.
2520 Peniche
Tel: 262 78 51 17 Fax: 262 78 71 71
socert@mail.telepac.pt
info@socert.pt
www.socert.pt

National associations for organic agriculture
Associação Portuguesa de Agricultura Biológica
Calcada da Tapara, 39 R/e dto
1300 Lisboa
Tel: 21 362 35 85 Fax: 21 362 35 86
agrobio@mai.teleweb.pt
www.agrobio.pt

Organic produce co-ops
BioCoop
Mercado Municipal Chão do Loureiro, loja 6
Lisboa
Tel: 21 886 05 95
Open Mon. to Fri. 10:00-14:00, Sat. 7:30-14:00

NaturoCoop
Bairro Fernando Magalhães, Bloco 13, Cave 2
Porto
Tel: 22 332 21 17
Open Sat. 10.00- 13.00

Cooperativa Terra Preservada
contact through BeirAmbiente
Vila Soeiro, Guarda
Tel: 271 22 49 00

Working on organic farms

WWOOF Independents
PO Box 2675
Lewes BN7 1RB
England, U.K.
www.wwoof.org
34 hosts listings in Portugal

Environment & nature conservation

Quercus
Associação Nacional de Conservação da Natureza
Apartado 4333
1503-003 Lisboa
Tel: 21 778 84 74 Tel/fax: 21 778 77 99
quercus@quercus.pt
www.quercus.pt

Solar cooking

It was our commitment to 'caring for life' (both the earth and humanity) that made us decide to move to the countryside near Mas Lluerna and to adopt an environmentally friendly lifestyle. We came to understand that our every deed uses energy, which in turn affects our surroundings. The growing hole in the ozone layer, caused by the overuse of energy, convinced us of the need for a renewable energy source that could meet all of our energy demands: electricity, water, heating... and our kitchen appliances!

We know that the sun is the energy source that sustains all life on Earth. People everywhere can use solar energy in one way or another. We have found a very pleasant and useful application of this source: our 'sun kitchen'. It not only saves energy but has also expanded our culinary possibilities. We can even preserve fruit and vegetables. Families from all over the region come to our cooking demonstrations as if they were parties. Using only the sun for heat, we cook the typical dishes of the Catalonia region. Our guests are always amazed that we use no fuels.

We use two types of sun kitchen:

1 - A box with a glass on top. You can build this with basic materials such as cardboard and wood. These kitchens rely on the principle of heat accumulation, trapping and retaining the heat from sun rays that shine in. The sun kitchens we have built from cardboard reach 150°C on sunny days.

In this type of kitchen, you can cook potatoes, puddings, grains, meat and fish in times ranging from 1 to 3 hours. The food is cooked slowly, which preserves flavour. Some foods, such as potatoes and marrow, can be cooked without water.

2 - Parabolic SK kitchens, as perfected by German scientist Dr. Ing. D. Seifert. These are made of galvanised aluminium and reflect the sun's rays to a central point, which is where you put your pan. This kitchen can reach a temperature of 350°C, and allows you to boil, deep fry, broil, pan fry, and even bake bread. Cooking times are the same as on an electric stove.

Sun kitchens emit approximately 0.5 kilos less CO_2 per hour than electrical or gas-fuelled cooking appliances. The Parabolic SK kitchen can be compared to a small power plant which uses solar energy directly - the cleanest, most efficient use of energy known to humankind.

Dr Seifert also started a programme linking European NGOs to NGOs in the South, with the aim of buying sun kitchens for all families who rely on wood for cooking. In this way, poor communities are supplied with simple, efficient technology that meets their energy requirements and reduces emissions of greenhouse gas.

More information about solar cooking: www.solarcooking.org

By Isel
20. Mas Lluerna, La Sentiu de Sió, Lleida, Catalunya

Farms in this guide that use a sun kitchen are: 20. Mas Lluerna in La Sentiu de Sió and 115. Cortijo de Garrido in Uleila del Campo.

Native breeds in Asturias

Native breeds of livestock are something of great value that we have inherited from our ancestors. During the 20th century, stock-breeding systems which give priority to productivity and profit became the norm. Considerations of quality, health, environmental respect and the animals' well-being were relatively neglected, leading to the virtual extinction of many local breeds, with the breeders concentrating on more 'profitable' ones. But how short-term will this profit be? The disappearance of a breed implies an irreparable loss, and also a wilful disregard for the fruits of hundreds of years of selection by our ancestors. We must preserve this inheritance for future generations.

In Asturias, the defence of indigenous breeds has for years been considered an 'aldeanada': old-fashioned, and related to backward, rural ways, whereas the importation of foreign races was seen as the 'modern' way to be followed. Fortunately there were always a few enthusiastic breeders who put great personal effort (and often scanty means) into the preservation of domestic Asturian breeds. Parallel to the worldwide phenomenon of conservation as a reaction to the abuses inherent in large-scale production techniques and an emphasis on purely economic objectives, there is a renewed

interest in native breeds. The Asturcón pony, the Roxa cow, the Xalda sheep, the Bermeya goat, the Pinta hen and an indigenous pig are the breeds native to the Principality of Asturias. Highly resistant to diseases and of excellent quality, these were the basis of Asturian domestic life for centuries.

Today, they are again becoming a fundamental part of Asturian life; living examples of its rich heritage and essential to the preservation of ecological balance. Here are three examples:

The Asturcón Pony

The Asturian Pony or Asturcón has become a symbol of the Principality of Asturias. Archaeological information from the seventh century BC demonstrates not only the existence of horses in Asturias, but also their domestication and use as working animals. The great value that the horse had for the native people in Roman times is well known, but when the Celtic peoples arrived, the importance of horses was greatly enhanced: they were the engine for Celtic expansion throughout Europe. The horse played an important role in their mythology: the veneration of Epona, a fertility goddess, probably started as the cult of an equine divinity. Representations of the goddess often show her sitting on horseback and surrounded by mares with their foals. During the Middle Ages, the horse retained its importance: Asturian peasants frequently paid the feudal lords' taxes with horses. Only recently, at the beginning of the 20th century, did the population of Asturcón Ponies began to decline.

The purity of the breed has been preserved through the ages. Together with the Exmoor, Shetland and Dartmoor Pony, the Asturcón is the most immediate descendant of the Celtic horse. The pony has a small and elegant build, with a long flowing mane. They naturally have an ambling gait, and as a result became popular as ladies' mounts. Nowadays this little horse, with its calm temperament and friendly character, is an excellent partner for children and a good starting point for learning to ride. There are currently around 800 Asturian ponies in the Principality.

The Xalda sheep

Like Brittany's Ousseant, the Welsh Black and Shetland's Morite, the Xalda of Asturias is a Celtic sheep. This is shown not only by archaeological finds, but also by quotes from classical authors, where 'Asturcensis' wool and black woollen tunics are mentioned. It is reckoned that there may well have been half a million Xalda sheep in the Principality of Asturias in the 18th century. This population began to drop half way through the 20th century, when most communal land was reforested with quick-growing eucalyptus and pine trees for the mining industry, using the pastures where the sheep had grazed for thousands of years. Also, great numbers of Xaldas were replaced by other breeds which were better producers of milk for cheese production. It was not until the 1990s that the importance of preserving this fine breed was recognized.

Xalda Sheep are agile, lively and good walkers. In general they look small and shapely. Their colour is black, white or a mixture of both. The future for the Xalda is now hopeful, as new breeders are being attracted; today there are about 550 sheep.

The Pita Pinta hen

The Asturian Pita Pinta is genetically related to the Atlantic branch of the species. With industrial avicultural development in the midst of the 20th century, and the creation of commercial hybrids in order to produce brown eggs, the indigenous Asturian hen practically became extinct. Only a few were left in remote places where the birds had never been totally commercialized. These breeders tried to raise the animals 'like their grandmothers did', taking a similar approach to that of modern conservationists. With the involvement of some enthusiastic sci-

entists, they managed to develop new groups of Pita Pintas. In the last few years these fine black and white birds with their characteristic mottled look have spread widely once again in the Asturian region.

By Severino García Gonzalez
53. Quintana de la Foncalada, Argüero (Villaviciosa)

At farm nº 53, Quintana de la Foncalada, you will find Ca' L'Asturcón, a museum which includes the history of the old Asturian and Celtic ponies and the Xalda Sheep.
At 49, La Montaña Mágica, you can go horse-riding on *Asturcónes* through the beautifull foothills of the Picos de Europa. Also, Xalda sheep are to be found on several Eceat farms in the region, and some keep small groups of the Pita Pinta hens.

	Tent	Caravan	Lodging	Self-catering	Hostel / group	Wheelchair friendly	Breakfast	Meals	Produce for sale	Swimming	Bikes to rent	Playground / toys	(…farm…)
SPAIN													
Catalunya													
1. El Molí, Siurana d'Empordà			•				•	•	•		•		
2. Can Bosc, Lladó				•	•		•	•					
3. Rectoría de La Miana, Sant Jaume de Llierca			•						•		•	•	•
4. Camping Masia Can Banal, Montagut	•	•				•	•	•	•	•	•	•	
5. Camping Rural Els Alous, Oix	•	•								•		•	
6. Mas Pujou, Olot						•				•	•	•	
7. Mas Cabrafiga, La Vall de Bianya			•					•	•			•	
8. Camping Mas la Bauma, Vallfogona de Ripollés	•						•	•	•	•		•	
9. Masia Serradell, Campdevànol				•	•		•	•	•	•		•	
10. Camping Molí Serradell, Campdevànol	•	•	•	•			•	•	•	•	•		
11. Camping Masia Can Fosses, Planoles	•	•					•		•			•	
12. Cal Pastor, Fornells de la Muntanya (Toses)				•	•		•	•	•				
13. Mas Can Sala, Sant Martí de Llémena	•		•				•	•	•			•	•
14. Masia Les Planes, Sant Mateu de Bages			•				•	•					
15. La Torre de Guialmons, Les Piles				•	•								
16. Mas de Caret, Montblanc	•		•	•			•	•	•	•		•	•
17. Mas de Mingall, El Perelló					•					•			
18. Casa Elisa & Can Torres, Masdenverge				•	•		•	•					
19. Venta de San Juan, Batea				•	•						•	•	
20. Mas Lluerna, La Sentiu de Sió	•	•	•					•	•	•	•		
21. Casa Pete y Lou, Tremp			•		→		•	•					
22. Cal Sodhi, Noves de Segre (Les Valls D'Aguilar)				•		•			•				
23. Camping Masia Bordes de Graus, Tavascán	•	•	•		•	•	•	•	•	•		•	
24. Casa Tonya, Unarre			•					•			•		
25. Camping Solau, Espot	•	•	•	•									
Aragón													
26. Casa Teixidó, Molins de Betesa			•				•	•					
27. El Rancho de Boca la Roca, Benabarre	•		•				•	•	•				
28. Allucant - Albergue Rural Ornitológico, Gallocanta			•		•	•	•	•	•		•	•	
29. Las Cardelinas, Pinsoro			•				•	•	•		•		
Navarra													
30. Primorena, Meoz			•				•	•	•	•			
31. Urruska Baserria, Elizondo			•				•	•	•				
32. Hotel Peruskenea, Beruete (Basaburua Mayor)			•			•	•	•	•			•	
33. Casa Etxeberri, Goldaratz				•		•	•	•	•				

	Tent	Caravan	Lodging	Self-catering	Hostel / group	Wheelchair friendly	Breakfast	Meals	Produce for sale	Swimming	Bikes to rent	Playground / toys	(Occasional) farmwork
Euskadi													
34. Aldarreta, Ataun			•				•	•				•	
35. Naera Haundi Baserria, Abaltzisketa				•			•	•	•		•		
36. Sarasola-Zahar, Aizarnazabal			•					•	•	•	•	•	•
37. Baserri Arruan Haundi, Deba	•	•		•	•	•	•		•				•
38. Mendiaxpe, Araia			•				•	•		•	•		
39. Erletxe, Laguardia (Biasteri)			•				•		•				
40. Uxarte, Aramaio (Ibarra)			•				•	•				•	•
41. Guikuri, Murua			•				•	•	•			•	•
42. Ibaizar Baserria, Sojo-Zollo			•	•			•	•	•		•		
43. Olabarrieta Beheko, Okondo			•				•	•	•				
44. Baserri Amalau, Zeanuri				•			•		•			•	•
45. Urresti, Gautegiz de Arteaga			•	•			•	•	•		•		
Cantabria													
46. Albergue Rural La Tejedora, Ojebar (Rasines)			•		•		•	•	•			•	
Asturias													
47. Agroturismo Muriances, Ribadedeva	•	•	•	•			•	•	•	•	•		
48. La Valleja, Peñamellera Alta			•			•	•	•	•				•
49. La Montaña Mágica, El Allende de Vibaño (Llanes)			•				•	•	•		•		•
50. El Correntíu, Ribadesella				•					•				
51. Posada del Valle, Collía (Arriondas)			•				•	•					
52. Posada Ecológica L'Ayalga, Piloña (Infiesto)			•				•	•	•				
53. La Quintana de la Foncalada, Argüero (Villaviciosa)			•	•			•	•	•	•	•	•	
54. La Casa del Naturalista, Argüero (Villaviciosa)			•				•	•	•		•		
55. La Llosa de Fombona, Fombona (Luanco)			•				•	•	•				
Galicia													
56. Camping Fragadeume, Monfero	•	•		•			•	•	•	•		•	
57. Casa Pousadoira, Callobre (Miño)			•				•	•	•	•			
58. Casa Paradela, Pobra de Trives			•				•				•		
Castilla y León													
59. Camping Brejeo, Vilela	•	•				•					•	•	
60. Albergue Las Amayuelas, Amayuelas de Abajo					•		•	•	•			•	•
61. El Linar, San Martín del Castañar				•			•	•					
62. El Burro Blanco, Miranda del Castañar	•	•											
63. La Lobera, Arenas de San Pedro			•			•	•	•			•		
Com. de Madrid													
64. Calumet, Berzosa del Lozoya				•	•	•	•	•	•		•	•	

	Tent	Caravan	Lodging	Self-catering	Hostel / group	Wheelchair friendly	Breakfast	Meals	Produce for sale	Swimming	Bikes to rent	Playground / toys	(Occasional) farmwork
Castilla-La Mancha													
65. El Descansillo, Escalera			•				•	•	•				
Com. Valenciana													
66. Mas d'Oncell & Mas del Rey, Catí			•					•					•
67. Mas de Noguera, Caudiel			•		•			•	•				•
68. La Surera, Almedíjar				•	•	•	•	•		•	•		
69. Permacultura Bétera, Bétera				•						•	•	•	
70. Casa del Río Mijares Albergue Rural, Buñol			•		•	•	•	•	•	•		•	•
71. Sierra Natura, Enguera	•	•	•	•		•	•	•	•	•	•	•	
72. Camping El Teularet, Enguera	•	•		•			•	•		•	•		
73. Mas de La Canaleta, La Serrella (Confrides)			•				•	•	•				•
74. La Higuera, Callosa d'En Sarrià	•	•		•			•	•		•			•
Islas Baleares													
75. Ets Abellons, Caimari			•				•	•		•			
76. Son Mayol, Felanitx			•	•			•	•		•			
77. Es Palmer, Colònia Sant Jordi			•				•	•	•	•	•		
78. Can Marti, San Joan de Labritja			•	•			•		•		•		
Extremadura													
79. Finca la Casería, Navaconcejo			•	•		•	•	•		•			
80. Finca Las Albercas - El Becerril, Acebo			•							•	•	•	
81. El Molino, Parador de Ti Mismo, Acebo			•										•
82. Caserío de Fuente de Arcada, Villamie			•	•		•	•	•		•			
83. La Huerta de Valdomingo, San Martín de Trevejo			•										
84. Finca El Cabezo, San Martín de Trevejo			•				•	•		•		•	
Andalucía													
85. El Cordonero & El Cedro, Fuenteheridos			•							•			
86. Las Navezuelas, Cazalla de la Sierra			•	•			•	•	•	•			
87. El Berrocal, Cazalla de la Sierra			•	•		•	•	•		•			•
88. Cañada de los Pájaros, La Puebla del Río			•										
89. Finca Paquita, Zahara de la Sierra			•							•			
90. Camping-Cortijo La Jaima, Prado del Rey	•	•	•							•	•		
91. Casa Montecote, Vejer de la Frontera			•			•				•	•	•	•
92. Casas Karen & Fuente del Madroño, Los Caños de Meca			•							•			
93. Casarosa, Estación de Gaucín-Colmenar				•			•	•					
94. Finca La Mohea, Genalguacil	•			•			•	•	•				
95. Finca Fantástico, Álora	•			•			•		•				
96. Casa de Elrond, Villanueva de la Concepción			•				•	•		•			
97. Las Cañadas, Mollina			•	•							•	•	

	Tent	Caravan	Lodging	Self-catering	Hostel / group	Wheelchair friendly	Breakfast	Meals	Produce for sale	Swimming	Bikes to rent	Playground / toys	(Occasional) farmwork
98. Finca Cuevas del Pino, Villarrubia				•					•	•			
99. Villa Matilde, Andújar	•		•				•	•		•	•		
100. Eco-turismo La Pendolera, Siles				•	•	•	•	•	•	•	•		
101. Las Castañetas, Villacarrillo				•					•	•	•	•	
102. Camping-Cortijo San Isicio, Cazorla				•					•				
103. El Cortijo del Pino, Albuñuelas				•			•	•		•	•		
104. Cortijo La Chicharra & La Chaparra, Lanjarón				•						•			
105. Cortijo La Lomilla, Orgiva				•			•	•	•	•			•
106. Cortijo La Torrera, Castell de Ferro	•		•				•	•		•	•		
107. Cortijo Buena Vista, Laroles				•	•		•	•		•	•		
108. Laveranda, Orce				•	•		•				•		
109. Cortijo Aloe Vera, Huércal-Overa				•	•		•	•		•	•		
110. Complejo Rural La Hierbabuena, Cuevas de Almanzora				•									
111. Cortijo El Nacimiento, Turre				•			•	•	•		•		
112. Finca El Rincon de Tablas, Turre				•									
113. Cortijo de Garrido, Sorbas	•		•				•	•	•		•		•
114. Cortijo Los Baños, Lucainena de las Torres				•		•	•	•	•	•		•	
Islas Canarias													
115. Montaña de Firgas, Firgas				•					•				
116. Atalaya de la Rosa del Taro, Puerto del Rosario				•									•
PORTUGAL													
The Beiras													
1. Quinta Casal da Fonte Grande, Trancoso	•	•		•					•	•			
2. Quinta da Lameira, Fornos de Algodres				•					•				
3. Quinta da Medroa, Guarda	•		•	•			•			•			
4. Quinta do Carvalhal, Guarda	•	•		•					•				•
5. Quinta do Ronfrio, Guarda (Trinta)	•			•			•						
6. Quinta da Alagoa, Guarda (Vale de Amoreira)	•					•	•	•	•			•	•
7. Quinta das Poldras, Seia (São Romão)				•									
8. Quinta Covão de Santa Maria, Manteigas	•		•				•	•	•				
9. Casa O Camponês, Gouveia (Nespereira)				•									
10. Casa da Fonte, Gouveia (Arcozelo da Serra)				•									
11. Quinta das Cegonhas, Melo	•	•				•	•	•		•		•	
12. Casa do Visconde, Canas de Senhorim	•	•	•				•	•	•	•		•	
13. Moinhos do Dão, Viseu (Alcafache)	•			•			•	•				•	
14. Quinta do Rio Dão, Mangualde	•	•		•			•		•			•	
15. Quinta da Chave Grande, Sátão				•					•		•		
16. Quinta da Comenda, São Pedro do Sul			•						•		•	•	

	Tent	Caravan	Lodging	Self-catering	Hostel / group	Wheelchair friendly	Breakfast	Meals	Produce for sale	Swimming	Bikes to rent	Playground / toys	(Occasional) farmwork
17. Campismo Natural da Fraguinha, S. Pedro do Sul	•	•		•			•	•		•	•		
18. Quinta da Cerca, Tábua (Casal da Senhora)	•	•	•				•	•	•				•
19. Quinta do Vale da Cabra, Oliveira do Hospital	•			•			•	•	•			•	•
20. Quinta das Mestras, Oliveira do Hospital				•			•						•
21. Residencial o Miradouro, Oliveira do Hospital	•		•						•				
22. Casa da Eira, Lousã (Casal de Ermio)				•			•		•				
23. Casa da Cerdeira, Lousã					•				•				
24. Quinta Vale de Asna, Góis (Vale de Asna)	•			•									
25. Quinta do Ribeiro, Arganil	•			•					•				
26. Quinta Regada, Arganil	•			•			•	•					
27. Quinta das Bouchas, Unhais da Serra (Paúl)	•	•								•			
28. Quinta da Cava, Unhais da Serra (Paúl)	•								•				
29. Quinta Vale Parais, Idanha-a-Nova	•			•			•	•	•				•
30. Camping PortUgo, Cernache do Bonjardim (Pampilhal)	•	•					•	•				•	
Estremadura, Ribatejo & Alentejo													
31. Campismo O Tamanco, Louriçal	•	•		•				•	•	•	•	•	
32. Casa da Maria Pequena & Casa Trindade, Bêco				•									
33. Camping Redondo, Tomar	•	•		•			•	•		•	•	•	
34. Campismo Rural da Silveira, Évora de Alcobaça	•	•										•	
35. A Colina Atlântica, Quinta das Maças, Barrantes			•						•	•			
36. Casal do Pomar, Caldas da Rainha				•									
37. Moinho de Vento, Alenquer (Penedos de Alenquer)				•									
38. Quinta da Picota, Bucelas				•	•		•	•	•				
39. Quinta do Pomarinho, Castelo de Vide	•	•	•							•	•		•
40. Quinta das Chaves, Évora				•			•	•					
41. Quinta Bica da Matinha, Cercal do Alentejo					•								
42. Quinta Cerca dos Sobreiros, Mina de São Domingos	•	•					•	•				•	
Algarve													
43. Olhos Negros, Monchique	•		•		•								
44. Quinta Vinha Velha, Lagos			•	•			•			•		•	
45. Monte Rosa, Lagos	•		•	•			•	•			•		•
46. Casa Stakwaas, Lagos (Barão de São João)				•									
47. Quinta Aloé Vera, Lagos (Barão de São Miguel)	•		•	•									•
48. Quinta do Coração, Loulé			•	•			•	•					

ECEAT
in Europe

ECEAT INTERNATIONAL
Chairman:
Fien Meiresonne, ECEAT NL
meiresonne12@zonnet.nl
General secretary:
Michal Burian, ECEAT CZ
michal@eceat.cz
Treasurer:
Matthias Baerens, ECEAT Deutschland
info@eceat.de
Board member:
Mirjam Olsthoorn, ECEAT Portugal,
beirambiente@mail.telepac.pt
Board member:
Niklas Palmcrantz, ECEAT Sweden
info@eceat.nu

BENELUX - Belgium, Netherlands, Luxembourg
contact person: Hans Owel
ECEAT NL
Postbus 10899
1001 EW Amsterdam, The Netherlands
tel: +31-20-668 10 30
fax: +31-20-463 05 94
eceat@antenna.nl
www.eceat.nl

Bulgaria
contact person: Pavlin Todorov
address: P.O. Box 134, 1233 Sofia, Bulgaria
tel: +359-2-59 49 25
mob: +359-88 61 35 59
pavlin@bulnet.bg

Czech Republic
contact person: Michal Burian
ECEAT CZ
Sumavska 31 b
61254 Brno, Czech Republic
tel: +420-5-413 250 80
fax: +420-5-413 2 50 80
info@eceat.cz
www.czechitnow.cz

Denmark
contact person: Mogens Godbale
Landsforeningen for Okologisk
Landboturisme / Danish National
Association for ecofarm-tourism
Brovej 7, 5464 Brenderup, Denmark
tel: +45-644 423 10
fax: +45-644 423 10
godballe@post12.tele.dk

Finland
contact person: Terhi Arell
Kaskenkatu 15a C15, 20700 Turku, Finland
tel: +358-2-251 83 01 or +358-40-519 49 49
terare@nic.fi

France
contact person: Auke van Hinte
ECEAT NL
Postbus 10899, 1001 EW Amsterdam,
The Netherlands
tel: +31-20-668 10 30 / +31-20-463 76 19
fax: +31-20 463 05 94
hinte1@zonnet.nl

Germany
contact person: Matthias Baerens
ECEAT Germany
Landreiterstraße 13, 19055 Schwerin
tel: +49-385-521 356 8
fax: +49-385-562 922
info@eceat.de
www.eceat.de

Greece
contact person: Michalis Probonas
"AEFOROS"
31 Kolokotroni str., 10562 Athens, Greece
tel: +30-10-324 73 64,
tel private: +30-810-39 32 81
fax: +30-10-322 43 44
aeforos@hol.gr

Great Britain & Ireland
contact person: Geert Snoeijer
ECEAT NL
Postbus 10899,
1001 EW Amsterdam, The Netherlands
tel 1: +31-20-668 10 30 / +31-20-463 76 24
fax: +31-20-4630594
g.snoeijer@antenna.nl
www.eceat.nl

Latvia
contact person: Peteris Locmelis
Peteris Locmelis, Audeju 7/9
LV 1966 Riga, Latvia
tel: +37-1-722 60 42
fax: +37-1-721 36 97
peteris@latviatours.lv

Norway
contact person: Cecilia Verheij
Cecilia Verheij, Mårtensbo Gård
PL 4281, 712 91 Hällefors, Sweden
tel/fax: +46-587-620 40
eceat@swipnet.se
www.eceatnorway.com

Poland
contact person: Jadwiga Lopata
ECEAT - Poland
34-146, Stryszow 156, Poland
tel: +48-33-879 71 14 / +48-33-879 78 16
fax: +48-33-8797114
jadwiga@eceat-pl.most.org.pl
http://sfo.pl/eceat/

Portugal
contact person: Mirjam Olsthoorn
ECEAT-Portugal - BeirAmbiente
Centro Profissional de Desenvolvimento
Sustentável e Eco Turismo
Vila Soeiro, 6300 Guarda, Portugal
tel/fax: +351-271-22 49 00
beirambiente@mail.telepac.pt
www.beirambiente.pt

Rumania
contact person: Zoltan Hajdu
FOCUS Eco-Center, O.P. 6, C.P. 620,
4300 Tirgu Mures, Romania
tel: +40-65-163692
fax: +40-65-163692
zhajdu@fx.ro

Spain
contact person: Esther Schasfoort
ECEAT NL
Postbus 10899,
1001 EW Amsterdam, The Netherlands
tel 1: +31-20-668 10 30
tel 2: +31-20-463 76 24
fax: +31-20-463 05 94
redactie.eceat@antenna.nl
www.eceat.nl

Sweden
contact person: Hans von Essen or
Niklas Palmcrantz
ECEAT Sverige
Idésmedjan Sjövägen 3, Hävla,
610 14 Rejmyre, Sweden
tel: +46-151-210 85 / +46-151-211 09
fax: +46-151-211 07
eceat@idesmedjan.com
tiseno@d.lrf.se
www.eceat.nu

Some environmental tips for your stay

We all look forward to relaxing and forgetting our worries, but even when on holiday we have a responsibility to care for the environment. Many tourist resorts have lost the natural beauty that attracted tourists in the first place! Therefore it is in all our interests that we preserve and improve the unique and attractive parts of our world.

Our biggest contribution lies in small things: our daily behaviour towards people and animals around us, to nature and the products that we use. The suggestions below will give you a few tips.

Save resources and respect the household rules of your hosts

- Your hosts have made significant efforts to bring your place to stay in harmony with the natural surroundings and to minimise their impact on the environment. If, for example, used goods are recycled as much as possible and waste is separated, put them in the appropriate bins.

- Take short showers, turn off taps after use, and report any water leaks

- Switch off unnecessary lights and other electrical appliances

- In order of preference: walk, rent a bike, use public transport, share taxis. If necessary, rent a motorbike or car with low fuel consumption. Use unleaded petrol if available.

Respect local culture

- Participate in and learn about local, social, religious and other customs

- Respect local dress codes, especially when sunbathing and when visiting religious venues

- Do not make excessive noise in residential areas or close to protected nature reserves

- Encourage rural economies by buying locally produced food, art and handicrafts rather than imported (tourist) goods

Protect nature

- Properly extinguish camp fires and cigarettes

- When walking, do not cut trees, flowers or plants

- Observe any restrictions and regulations by local authorities

- Do not buy products made from protected animal or plant species

- Do not leave litter on land, nor in the sea. Plastic and metal do not biodegrade, and will remain as an eyesore for many years.

Support local environmental activities

- Write to tourist boards and newspapers in your host country to express your views, and send photographs to reinforce your message. As a consumer, your opinion carries a lot of weight.

- Many countries suffer from excessive hunting of (protected) animals, birds and fish. Write to your local press, expressing your objections to these practices.

- Hotels and tour operators should be encouraged to adopt policies to reduce the environmental impact caused by tourism. Ask them to make environmentally-sound improvements.

- Local authorities should keep a balance between encouraging local industry and protection of the environment. Write to them with suggestions.

Parts of the above have been derived from the code set up by the Mediterranean network of Friends of the Earth, financed by the European Commission.

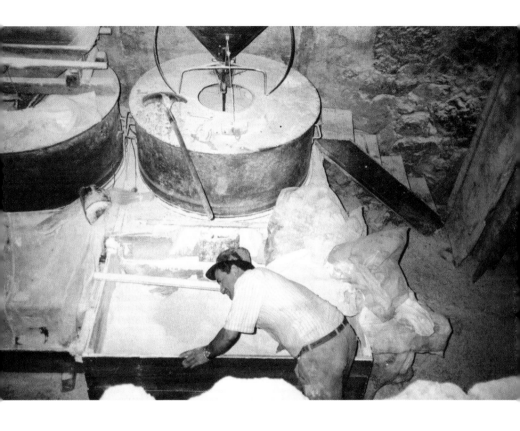

Quality control

A holiday on an organic farm is intended to support the development of organic farming. Farmers make great efforts to produce food in as sustainable a manner as possible, and to protect the natural environment. By spending time in such authentic and agreeable natural surroundings, holidaymakers show the farmers that their work is appreciated, and it also gives the farms extra income, which helps them to realise their goals.

ECEAT selects its places to stay in different ways, according to national standards. In Germany for example, only organic farms are eligible for entry, whereas in England and Poland the focus is wider, including smallholders and especially environment-friendly venues in national parks or other areas which are part of the cultural heritage.

For Spain and Portugal, the questionnaire includes the following issues:

Organic land and garden management

- Abstention from the use of pesticides and artificial fertilisers
- Animal welfare: space and hygiene
- Non-intensive land use
- Rare breeds or crops are kept/grown
- Natural elements are protected and actively restored (dry stone walls, hedgerows, hay meadows, native trees/bushes, wetlands etc.)
- Active enhancement of wildlife (bird boxes, ponds etc.)
- Where working farms are accredited with certifying bodies, the organic management is not checked in depth, as this has already been done by the certifying body.

Ecological household

- Products used are bought locally
- If food is served, it is organic as much as possible
- Use of biodegradable cleaning products

Avoidance of waste

- Minimal use of pre-packed food and one-person packages
- Waste separation and storage

Water and energy

- Low-flow showers and taps
- Reduced water usage for toilets and recycling of greywater
- Wetland or reed bed sewage treatment
- Low-energy light bulbs and automatic switch-offs
- Insulated walls, double glazing and individual heating controls
- Use of solar, hydraulic, biomass or wind energy

Building materials

- Environment-friendly paints, traditional building materials and wood preservatives

Situation of place to stay and hosts

- Situated within or close to protected areas of special interest
- A range of activities on the spot and nearby
- Hosts care for the needs of their guests, are friendly and helpful in passing on information about their premises, natural surroundings, wildlife and nearby attractions. Hosts are actively involved in social and rural development at a local level.

Hygiene and safety for children

- Farms are fantastic places for children, but it should be kept in mind that visiting a working farm is evidently different from staying in a four-star city hotel. Hygiene and safety are important issues, and are checked and described as accurately as possible. It is advisable to announce in advance that you will be staying with children, and to ask whether children are free to roam around and, for example, pet the sheep or play with the horses.

Accessibility

- In some instances, farms are not able to receive cars with campers or caravans. In that case, prices are given for tents only. Note that some farms will not allow cars on the campsite next to the tents, in order to preserve to the natural character of their land. A few places have installed facilities specifically geared to wheelchair-users, which is mentioned in the text. For other venues, please check in advance with your hosts.

Evaluation

Name and address of place where you stayed   Nr in guide:

Date of arrival   Date of departure

How many people were in your group?   Ages:

How did you find about The Green Holiday Guide?

☐ **I enjoyed The Green Holiday Spain & Portugal because**

☐ **I did not like The Green Holiday Guide Spain & Portugal because**

The descriptions of the places to stay in the guide were

☐ very accurate   ☐ satisfactory   ☐ bad

Comments

Comments on the place to stay:

☐ camping   ☐ lodging   ☐ self-catering   ☐ hostel

☐ other:

Evaluation

What did you think of the attractivity, character and position of the place to stay

☐ superb ☐ allright ☐ disappointing

Comments

What did you think of the attractivity, character and position of the immediate surroundings

☐ superb ☐ allright ☐ disappointing

Comments

Is it an attractive place for children and is it safe enough for them?

☐ superb ☐ allright ☐ disappointing

Comments

What did you think of the hygienic conditions of

Sanitation

☐ very good ☐ satisfactory ☐ insufficient ☐ bad

Kitchen

☐ good ☐ satisfactory ☐ insufficient ☐ bad

Living area and bedroom

☐ good ☐ satisfactory ☐ insufficient ☐ bad

Comments

Were the hosts hospitable?

☐ yes ☐ fairly ☐ no

Comments

_____

_____

How would you assess the standard of comfort of the services offered on the spot

☐ high   ☐ average   ☐ below average   ☐ basic

Comments

_____

_____

How would you assess the environmental soundness (issues such as eco-friendly cleaning products, waste separation and recycling, energy-saving measures, organic food – if applicable etc.)

☐ high   ☐ average   ☐ below average   ☐ bad

Comments

_____

_____

How would you qualify the choice of activities available on the spot and in the immediate surroundings

☐ plenty   ☐ just enough   ☐ not enough

Comments

_____

_____

**I know an organic farm which would like to be included in the next edition.
Its address is:**

_____

_____

_____

Which interesting place(s) to visit is/are worth mentioning in the next edition?

_____

_____

Supporting ECEAT

ECEAT is a non-profit association with the aim to enhance small-scale, sustainable tourism on organic farms and other environment-friendly holiday destinations on the European countryside. With your support the quality and the range of our guidebooks can improve as well as our internet services and other promotional activities like brochures and travel fairs. From € 14 per year you can be a supporter of ECEAT and receive 30% discount on the travelguides as well as our newsletter.

☐ yes, I will support ECEAT and I include a cheque of

€ _____

made to ECEAT

Signature:

_____

If you like, you can send us some of your holiday pictures which may come out in a future edition (please write name of the place to stay and country on the back of the pictures).

Thank you for completing this form.

Name:

_____

Address:

_____

_____

_____

Send this form or a photocopy to

ECEAT
C/o Holiday Evaluations
P.O. Box 10899
1001 EW Amsterdam
The Netherlands

Order form Green Holiday Guides

Please mark the guide you would like to receive below.

Guides available in 2002:

In English
- ☐ **Poland** *(€7.95)*
- ☐ **Czech Republic** *(€7.95)*
- ☐ **Great Britain and Ireland** *(€11.75)*
- ☐ **Nordic Countries** *(Norway, Sweden, Finland, Denmark; Jan. 2003)*
- ☐ **Spain and Portugal** *(€11.75)*

In German and/or Dutch
- ☐ **Germany** *(German, €9.75)*
- ☐ **France** *(Dutch, €7.95)*
- ☐ **Italy** *(German, €9.75)*
- ☐ **Switzerland** *(German, €7.95)*
- ☐ **Austria** *(German, postage only)*
- ☐ **Europe** *(Dutch, €15.95)*

Postage is €3

Name

Street

Town

County

Country

Tel

Email:

Date   Signature

I enclose a €   cheque including the €3 postage costs,
or the equivalent in UK £.

Send this form or a photocopy to

ECEAT
C/o Customers' Service
P.O. Box 10899
1001 EW Amsterdam
The Netherlands

Order form

Also available from

CREATING A FLOWER MEADOW

Yvette Verner
"An inspiring story of our times"-David Bellamy

Shows how you can create a natural habitat in your own garden, whether large or small, so you too can observe the lifestyles of our native flora and fauna, and play your part in encouraging their survival. Includes lists of species of flowers, grasses, trees, birds and butterflies, a seasonal calendar and contact addresses for further info.
Green Earth Books 144pp with 40 line drawings and 16pp of colour plates, bibliography and index 234 x 156mm ISBN 1 900322 08 0 £9.95 pb

THE ORGANIC DIRECTORY

2002/3 edition
Compiled & edited by Clive Litchfield

"A very handy guide for foodies and greenies both"-Food Magazine

The Organic Directory is the most comprehensive listing of its kind. Arranged on a county-by-county basis to help you buy locally, this new edition covers England, Scotland, Wales, Northern Ireland and the Channel Islands. In The Organic Directory you will find the names, addresses and phone numbers of: • Retailers and producers of organic food • Vegetable box schemes (weekly boxes of in-season vegetables from organic farmers) • Suppliers of organic gardening materials • Restaurants and accommodation specializing in organic food • and a wealth of other information including details of labelling schemes for organic produce; farm shops and farm gate sales; the WWOOF (willing workers on organic farms) movement; and education opportunities.
Published by Green Books with the Soil Association 384pp with index 190 x 120mm ISBN 1 903998 10 7 £4.95 pb

GO M.A.D.

(Go Make A Difference)
365 Daily Ways to Save the Planet
Compiled and edited by The Ecologist

The mainstream media likes to pretend that living a more ecologically responsible and sustainable life is either more expensive, or more difficult than contemporary culture can allow. The Ecologist's editors disagree fundamentally. They believe that a more ecologically responsible and sustainable lifestyle is not only less expensive in most regards but also more enjoyable and above all happier! To this end they've compiled this pocket guide packed with 365 practical and useful tips on how we can all take daily actions that will benefit our environment, ecology and community. The tips have been collated with the kind help of hundreds of excellent ethically aware companies, organizations, movements and charities. And for each of the tips, there's a straightforward reason as to why they are so important. So there really are no reasons for not adopting each and every tip. Quite simply Go Make A Difference seeks to narrow the gap between the principles and practices of ecological living. So what are you waiting for, Go Make A Difference!
Published by Think Publishing on behalf of The Ecologist, distributed by Green Books 180pp 178 x 111mm ISBN 0 9541363 0 6 £3.99 pb

THE NEIGHBOURHOOD FORAGER

A Guide for the Wild Food Gourmet
Robert K. Henderson

Organized into chapters based on plant type (evergreens, broadleaf trees, common flowers, peripherals, and plants with edible greens and roots), The Neighborhood Forager introduces the aspiring wild-gatherer to more than 60 genera of plants, comprising hundreds of widespread species. Full of history and folklore, the book offers a wide range of practical and entertaining information.
Chelsea Green (USA) 240pp with 8 page colour section and 100 illust. & photos, index 234 x 190mm ISBN 1 890132 35 7 £16.95 pb